MEASURING SUSTAINABLE DEVELOPMENT GOALS PERFORMANCE

MEASURING SUSTAINABLE DEVELOPMENT GOALS PERFORMANCE

STEN THORE
University of Texas, United States

RUZANNA TARVERDYAN
The Geneva Consensus Foundation, Switzerland

ELSEVIER

Elsevier
Radarweg 29, PO Box 211, 1000 AE Amsterdam, Netherlands
The Boulevard, Langford Lane, Kidlington, Oxford OX5 1GB, United Kingdom
50 Hampshire Street, 5th Floor, Cambridge, MA 02139, United States

Library of Congress Cataloging-in-Publication Data
A catalog record for this book is available from the Library of Congress

British Library Cataloguing-in-Publication Data
A catalogue record for this book is available from the British Library

ISBN: 978-0-323-90268-7

For information on all Elsevier publications visit our website at https://www.elsevier.com/books-and-journals

Publisher: Joe Hayton
Acquisitions Editor: Kathryn Eryilmaz
Editorial Project Manager: Michelle Fisher
Production Project Manager: Maria Bernard
Cover Designer: Miles Hitchen

Typeset by TNQ Technologies

Contents

Foreword

In today's rapidly changing world, the importance of rigorous analysis to inform policy making using publicly available data cannot be underestimated. The United Nations' 2030 Agenda for Sustainable Development, with its 17 goals, 169 targets, and 230 indicators, provides a rich framework to analyze the performance of countries over time and to compare to their peers. The people-centered and universal goals and targets defined in the 2030 Agenda are integrated and indivisible, and they balance the three dimensions of sustainability—economic development, social progress, and environmental protection—simultaneously seeking to leave no one behind and to strengthen universal peace. This publication uses open data sources and well-established statistical methods to offer the reader a set of conclusions that are evidence-based and transparently derived, thus providing us with unbiased insights.

The 2030 Agenda unanimously agreed by 193 heads of states is a powerful manifestation of commitment to the future we want and of political will to forge a global partnership by all countries and stakeholders to attain the Sustainable Development Goals (SDGs). These transformative SDGs, defined in this new historical plan of action for people, planet, and prosperity, are an articulation of solidarity. This is precisely what is needed in times of great despair and uncertainty, partly due to the ongoing pandemic but also to continuing technological transformation and globalization, which leave many of our citizens with a sense of insecurity about whether or not living standards will continue to improve.

It cannot be emphasized enough that the finish line of 2030 is fast approaching, and analyses like the one presented in this book can assist to speed up the attainment of the SDGs. In 2019, the UN Secretary General called on all sectors of society to mobilize for a decade of action to generate an unstoppable movement pushing for the required transformations needed to implement the 2030 Agenda. However, it is clear to many experts in the field that the world is not on track in achieving the SDGs.

Put simply, data tell us where we stand in the race to achieve the SDGs. Ratings and rankings can always be further improved, but metrics can be a great motivator and provide a powerful challenge to the policymaker—the same way they motivate an Olympic athlete to improve the focus of his or her training and the use of available resources to go "faster, higher, and stronger." This publication seeks to convey the richness of the statistics sets using a data-driven method to tease out insights for country policy makers and development practitioners that can assist them to improve the performance of their country on the attainment of the SDGs.

The policy effectiveness rating proposed by the authors transcends the standard measures of economic, social, and environmental performance. It points out to various inefficiencies and development challenges and distills lessons that governments and other development organizations and partners alike can take into account when developing comprehensive policy advice for regional, industry and global agendas while we strive to implement the 2030 Agenda for Sustainable Development.

Mahmoud Mohieldin
UN Special Envoy for the 2030 Development Agenda,
Executive Director International Monetary Fund

Preface

Our warmest thanks are due to the senior management at Elsevier for inviting us to expand our 2016 article "The Sustainable Competitiveness of Nations," *Journal of Technological Development and Social Change* into a full-length book and for having patience with us as our manuscript gradually evolved.

The present result should be seen as an update and sequel to our 2015 book *Diagnostics for a Globalized World,* Now Inc and World Scientific. Just a few months after that earlier work rolled off the printing press, world leaders at a UN meeting in New York agreed on their *2030 Agenda* for world sustainable development, listing 17 sustainable development goals. The *Agenda* urged all nations to coordinate their social and economic policies aiming for these goals—the very kind of task that we had addressed, providing the mathematical machinery for assessing the success or shortcomings of the performance of each nation.

The arrival of the COVID-19 pandemic has made social and economic policy infinitely more difficult. The attempts of nations to control the spread of the virus and to contain the death rates no doubt impact in many as yet unknown ways on their sustainability record. We shall learn more about this as UN member nations file their reports at the Statistical Commission.

We ask the indulgence of our readers in understanding that our present text is thus very much time defined, with the numerical calculations reported in Part I being concluded in early December 2020 and the text in Part II reflecting conditions and events up to the spring of 2021.

Our sincere thanks to Michelle Fisher, Elsevier, who good-heartedly found order in our scattered pieces of manuscript and brought them all together to a final book and also to Maria Bernard who expertly piloted the proofs through the intricacies of the Elsevier computerized submission system.

S.Th., The Algarve, Portugal
R.T., Geneva, Switzerland

Introduction

In our earlier book "*Diagnostics for a Globalized World*" (Thore and Tarverdyan, 2015), we lamented the growing disconnect between the problems and tasks of the real world and the subject matters dealt with in economic textbooks. Six years later, the economic health of the globe is waning in the face of severe famine in many parts of the developing world, increasing warfare in Asia, South Caucuses, and Africa, and a world-wide virus pandemic.

The prevalence of poverty among plenty, the need amidst unused resources, and the destitution on the footpaths of riches are not new problems on the agenda of humanity. David Beasley, head of the World Food Program based in Rome stated in April 2020 "we could be facing multiple famines of biblical proportions within a short few months." And the director of the Food and Agricultural Organization noting that widespread conflict and instability has led to food insecurity, said "We have mobilized our organizations in ways not seen since the foundation of the UN" (Reported in UN News, April 2020).

In this new world, economics needs to deal with persistent hunger, widespread disease, growing inequality, the mounting environmental degradation, and devastating climate change. To deal with these challenges, economic theory needs to understand markets in disequilibrium, disorder, even chaos. The COVID-19 necessitates a complete rethinking of received social and economic theory and a reordering of national priorities in the light of the UN sustainable development goals (SDGs).

Recent initiatives

The concept of Decent Work, as promoted by the International Labour Organization since 1999, encompassing job creation, rights at work, social protection, and social dialogue (with gender equality as a cross-cutting objective), quickly became reinterpreted in terms of the millennium goals. Similarly, the idea of "A Fair Globalization" was full-heartedly embraced by the international community, putting the productive employment and decent work at the heart of policy making for achieving a fair globalization and poverty reduction.

Adopted during the largest-ever gathering of world leaders—a three day summit held on 6–8 September at New York; the "United Nations Millennium Declaration" (UN, 2000b) was a major milestone in spelling out priorities and goals for international economic and social policy. It stated values, principles, and objectives for the international agenda for the 21st century and set deadlines for collective action. Recognizing that "only through broad and sustained efforts to create a shared future, based upon our common humanity in all its diversity, can globalization be made fully inclusive and equitable," world leaders unanimously declared that the central challenge of the day was to ensure that globalization becomes a positive force for all, acknowledging that both its benefits and its costs were unequally shared. The Declaration called for global policies and measures corresponding to the needs of developing countries and economies in transition.

The outcome document of the United Nations Conference on Sustainable Development—or Rio+20 Conference, "The Future We Want," adopted in Rio de Janeiro in June 2012 contains clear and practical measures for implementing sustainable development and laid the foundation for a new era of partnership. In Rio, the member states decided to launch a process of developing a set of SDGs, building upon the Millennium Development Goals.

Then, in September 2015 came the clarion call to action by world leaders assembled in New York: during the United Nations General Assembly, by resolution 70/1 of 25 September 2015, 193 heads of the UN Member states unanimously agreed on a global *2030* Agenda *for Sustainable Development* (UN Agenda 2030, 2015). The Agenda is a plan of action for people, planet, and prosperity—listing a set of SDGs and spelling out coordinated directions for international economic, social, and environmental policy.

The Agenda 2030 presents 17 SDGs, for developing and developed countries alike, each one in its turn characterized by a large number of "indicators." The goals provide a blueprint for the transition to a healthier planet and a more just world for the present and future generations. With concrete targets, the goals aim to end poverty and hunger, expand access to health, education, justice, and jobs, and promote inclusive and sustained economic growth, while protecting our planet from environmental degradation.

By its resolution 69/313 of 27 July 2015, the General Assembly endorsed the Addis Ababa Action Agenda of the Third International Conference on Financing for Development, seeking to develop high-quality, timely, and reliable data disaggregated by sex, age, geography, income, race, ethnicity, migratory status, disability, and other characteristics relevant in national contexts.

The light of inclusive multilateralism further shone brightly with the adoption of two significant global agreements: the Sendai Framework for Disaster Risk Reduction 2015–2020 adopted at the Third UN World Conference in Sendai, Japan, on March 18, 2015 and the Paris Agreement under the United Nations Framework Convention on Climate Change that entered into force on November 4, 2016.

In response to the urgent call for a data revolution, in 2015, the United Nations Statistical Commission established the High-level Group for Partnership, Coordination, and Capacity-Building for statistics for the 2030 Agenda for Sustainable Development (HLG-PCCB) and tasked the Group to provide strategic leadership for the SDG implementation process as it concerns statistical monitoring and reporting.

The global indicators framework for SDGs was developed by the Inter-Agency and Expert Group on SDG Indicators and agreed upon at the 48th session of the United Nations Statistical Commission held in March 2017. The global indicator framework was later adopted on July 6, 2017 by the General Assembly Resolution on Work of the Statistical Commission pertaining to the 2030 Agenda for Sustainable Development (A/RES/71/313). The framework includes 231 unique indicators (twelve indicators repeating under two or three different targets make the total number 247).

According to the Resolution, the indicator framework will be refined annually; it was reviewed comprehensively by the Statistical Commission at its 51st session in March 2020 and its 56th session, to be held in 2025. The global indicator framework will be complemented by indicators at the regional and national levels, which will be developed by Member States.

The Cape Town Global Action Plan for Sustainable Development Data was launched at the first United Nations World Data Forum, held in Cape Town, South Africa, from 15 to 18 January 2017, and endorsed by the Statistical Commission at its 48th session and which provides the framework for discussion, planning, implementation, and evaluation of statistical capacity-building pertaining to the 2030 Agenda for Sustainable Development.

Several international agencies are now involved in creating data banks of such indicators. All stakeholders: UN members, the United Nations funds and programs, the specialized agencies, the secretariat including the regional commissions, the Bretton Woods institutions, are to increase their efforts in data collection and statistical capacity-building, strengthening the coordination among national statistical offices as appropriate within their mandates. Similarly, the Data For Now (Data4Now) initiative was launched by the Deputy Secretary-General. It is organized by the United Nations, the Global Partnership for Sustainable Development Data, the World Bank, and the Sustainable Development Solutions Network. It will build new partnerships of governments and national statistical agencies to deliver robust and timely data helping users, leaders, and innovators to achieve the SDGs.

The UN Statistics Division in 2019 produced a series of reports and publications reviewing progress towards the achievement of the SDGs, including:

The Sustainable Development Goals Report 2019;

The Sustainable Development Goals Progress Chart 2019;

Progress on the Sustainable Development Goals: The Gender Snapshot 2019; and

The report of the Secretary General entitled "Special edition: progress towards the Sustainable Development Goals" (E/2019/68).

The Sustainable Development Goals Report 2019 is an illustrated version of the progress report and targets a wider audience interested in progress on the goals by presenting data and analysis in a user-friendly way with maps and charts. It includes an overview section with infographics highlighting the most important achievements and the most significant gaps and challenges to date for all 17 goals.

The UN statistics division publishes annual "progress charts," measuring the progress or lack of progress of nations towards the SDGs. In the Progress Chart for 2019, the results were summarized at regional levels for each of the 17 SDGs. In sub-Saharan Africa, there was "very high poverty," "very high undernourishment," "very high" incidence of malaria and HIV, "low proficiency" of reading at the end of primary education, "high rate of child marriage" for girls, and so on, down the entire list. Similarly, the UN Food and Health Organization recently released a report documenting that hunger in Africa continues to rise (Africa Regional Overview of Food Security and Nutrition, 2019).

The Sustainable Development Goals Progress Chart 2019 presents a snapshot of progress made towards selected targets under all goals. It helps the reader to visualize global and regional progress towards the achievement of the goals using traffic light colors to indicate the levels of progress, based on some of the indicators and information available as at September 2019.

The progress chart was included in a summary edition of The Sustainable Development Goals Report 2019 and circulated at the Sustainable Development Goals Summit in September 2019.

Progress on the Sustainable Development Goals: The Gender Snapshot 2019 was prepared jointly by the United Nations Entity for Gender Equality and the Empowerment of Women (UN-Women) and the Statistics Division. It brings together the latest available evidence on gender equality across all 17 goals. It highlights areas across all the goals in which women and girls remain disadvantaged, noting where further efforts are needed to ensure that no one is left behind. It also identifies gender-specific indicators of the global indicator framework used to measure the experiences of women and girls in the fulfillment of the goals.

As admitted in the Report of the Secretary-General, entitled *"Special edition: progress towards the Sustainable Development Goals,"* E/2019/68, to High-level segment: ministerial meeting of the high-level political forum on sustainable development, convened under the auspices of the Economic and Social Council conducted from 26 July 2018—24 July 2019:

> *"there is no escaping the fact that the global landscape for Sustainable Development Goal implementation has generally deteriorated since 2015, hindering the efforts of Governments and other partners. Moreover, the commitment to multilateral cooperation, so central to implementing major global agreements, is now under pressure."*

The report identified a series of cross-cutting areas where political leadership and urgent, scalable multistakeholder action is needed for the next decade of implementation to dramatically accelerate progress. Hopefully, this will enable the UN to shift the world onto a trajectory that is compatible with the achievement of the SDGs by 2030.

The list of challenges for sustainable developments facing the global community is a staggering one:

- reaffirming the pledge that no one will be left behind in implementing the 2030 Agenda for Sustainable Development and that the Agenda is people-centered, universal, and transformative;
- that the SDGs are integrated and indivisible, balancing economic, social, and environmental dimensions;
- developing a plan of action for people and planet strengthening universal peace and freedom, to be implemented by all countries and stakeholders in collaborative partnership;
- eradicating poverty in all its forms and dimensions, including extreme poverty.

Analytical treatment

Delivering on the UN agenda and continuing our work on measuring the performance of multidimensional concepts, the purpose of the present volume is to discuss the successes (or failures) of the social, environmental, and economic policy of nations aiming for the attainment of the UN SDGs. One of those goals (SDG VII) is to rid the world from pandemics, like the HIV and the COVID-19 virus. Progress is being made against diseases such as hepatitis, where the incidence of new chronic hepatitis B virus infections has been reduced considerably. As is now abundantly clear, however, the virus is not just a matter of health. It is also a matter of coordinating conflicting medical advice and conflicting political ambitions. Nations have met this challenge in different ways, often spurning the need for global consultation and coordination (SDG XVII), instead leaving us with a bewildering array of exhibits of alternative policies.

In publications spanning the last 20 years, the present authors have discussed such multidimensional concepts and shown how it is possible to determine the weight of each participating factor endogenously, by mathematical programming solving multiple problems—one for each individual country. We were pleased to learn in the Official Journal of the European Union—"Opinion of the European Economic and Social Committee on Listening to the citizens of Europe for a sustainable future (Sibiu and beyond) (2019/C 228/06)" adopted at plenary session No 542, that our 2016 study of Sustainable Competitiveness drew the attention of the Economic and Social Committee of the European Union (Thore and Tarverdyan, 2016, EESC 2019). With 147 votes against 64, the Committee decided to use our definition of "Sustainable Competitiveness," rather than the fixed-weights index published by the World Economic Forum, as our model balances economic prosperity, environmental issues, and social inclusiveness. The Committee concluded that the sustainability-adjusted global competitiveness index needs to take into consideration two new dimensions—both environmental and social ones.

In a chapter published in the "Decent Work and Economic Growth" volume of the Springer Encyclopedia of the UN SDGs, entitled "Beyond GDP: Saving the Planet by Measuring the Effectiveness of Policies" we offer an analytical framework for examining the interlinkages between economic growth, social equity, and sustainable development (Springer 2021).

The COVID-19 challenge and the Great Reset

United Nations Secretary-General António Guterres, in his briefing to the Security Council on Global Governance Post-COVID-19, stated that "the COVID-19 pandemic is one of the most dangerous challenges this world has faced in our lifetime."

The loss of lives and the economic destruction severely impacts on the sustainability achievements of nations. Suddenly one single goal of economic policy is gaining precedence to others: to contain the spread and incidence of the virus. Already it is becoming clear that combatting the virus requires more than doctors and carers and hospital equipment. Access to clean drinking water, sanitation, and a healthy life style are just as important. Protective barriers have to be created, like social distancing, lockdowns, or even the closure of entire regions. In response to the onslaught of the virus, many nations are searching for novel models of conducting business, like on-line shopping, closing downtown office buildings, and fossil-free transportation. The virus is wreaking havoc with established policy routines as we know them. Eventually, in years to come, the picture will settle and the net effects on the SDG records of each country will be documented.

In the meanwhile, our modest contribution is to offer an analytical format that identifies countries reaching optimal COVID-19 survival results—optimal as part of an optimal package of other important SDGs. For a country falling behind this optimum, we identify its shortcomings (its deficit COVID-19 survival rate and other goal deficits).

Sustainable development will continue to be the most significant collective challenge facing humanity for the years to come. Addressing it without compromising our shared prosperity and the right to development of future generations would require a significant alteration of macroeconomic governance. It calls for strategic vision and innovative policy responses. Enhanced data availability will create the information basis for development of global models, thus providing an analytical underpinning for global governance. This book examines the promise that is offered in that respect by the operations research technique called Data Envelopment Analysis, but the large scale practical application is still awaiting.

PART I MULTI-DIMENSIONAL ANALYSIS

BY STEN THORE

Overview

The Agenda 2030 calls for integration of the multiple dimensions of sustainable development into a single framework; there is a wide consensus on the need for a multi-disciplinary analysis embracing phenomena such as globalization, migration and the climate that impact upon the sustainable development goals (SDGs) will always be multidimensional. Once we have agreed on a single numerical index of the indicators, it becomes possible to document numerically how policymakers succeed (or fail) in achieving their goals. Statisticians, however, conventionally have had difficulties facing this problem. The unfortunate practice of forming such an index by setting fixed and given a priori weights to each factor (such as determining them by "experts" or simply giving each factor equal weight) is patently unrealistic. Each nation faces its own problems and will necessarily have its own priorities.

Our analysis of these matters follows the lead of the Dutch economist Jan Tinbergen, shared recipient of the very first Nobel prize in economics, who in the 1950s wrote two standard works on economic policy (Tinbergen 1952, 1956). Tinbergen saw the task of economic and social policy as a problem of mathematical optimization, maximizing a nation's "welfare function," subject to constraints set by technology, resources, and (to some extent) political feasibility. Following his lead, we shall calculate an index of SDGs for each nation as an instance of a "SDG welfare function." Again a multidimensional concept, this time with no less than 17 dimensions.

We define the effectiveness of the SDG policy of a nation as its ability to convert its policy instruments into SDG welfare. Assume that each nation tries to maximize the effectiveness of its policy. Ranking nations in terms of their SDG achievements, data show that a few nations are able to convert their policy instruments into maximal returns of the goals. We will be looking for champion nations that are superior to the others in the sense that there exists no other nation producing the same goals but using less or the same policy instruments. As we shall see, they will typically include both rich and some quite poor nations. They are ascribed an effectiveness rating equal to 1. Taken together, the champions define the maximally achievable SDG results. A nation falling short of champion performance experiences goal deficits, signaling that its policy efforts are lacking to deliver the desired development outcomes.

An easy way of thinking of these things is in terms of a frontier—a battle frontier drawn between the achievable and the nonachievable, between the doable and the nondoable. A few champion nations reach maximal SDG results. They mark the strongholds of the frontier. Other nations fall behind the frontier. They are rated as subeffective, with a policy effectiveness rating less than 1. Their actual goal achievements fall short of the maximally possible derived from the empirical welfare functions. To the mainstream economist, this is a distressing thought: economic policy producing results that are less than the maximally achievable. It flouts the accepted dogmas of general equilibrium, so dear to earlier generations of economists.

How is it possible that some nations, while still seeking the maximal effectiveness of their policy efforts yet obtain only suboptimal results? The answer is that such nations are applying an inappropriate or unfortunate welfare function, leading them to adopt policies that are globally suboptimal. There is only one way to solve this dilemma: revisiting the national goal priorities and thus recalibrating the policies themselves.

The recognition of suboptimal behavior involves more than just a willingness to tolerate some short-term deviations from the equilibrium world. As we see it, it indicates the need to rewrite the entire neoclassical economic system, just as developments in physics has led to a reformulation of the Newtonian system. The Nobel laureate Ilya Prigogine has proposed such a reformulation, based on the idea that most physical systems are located far from equilibrium and stay in disequilibrium most of the time. Similarly, our numerical work has led us to believe that the achievements of economic and social planning in many countries in the world are located far from equilibrium, falling short of optimum.

CHAPTER ONE

Economic growth and sustainability*

1.1 Gross domestic product and beyond

In this first chapter of the present volume, we investigate the causal relations between the gross domestic product (GDP) of a nation and its "sustainable development goal (SDG) welfare function," written as a weighted index of its SDG achievements. The text addresses readers with an interest in international relations, foreign trade, and the social and economic development of nations. Technical matters are avoided or dealt with rather summarily; they will instead be discussed in Chapter 2 of the book.

As long as economists have calculated the GDP of a nation, there has been the recognition that it does not measure societal progress beyond pure economic and financial indicators. In his 1971 Nobel memorial lecture, Simon Kuznets stressed that "the welfare of a nation can scarcely be inferred from a measurement of national income." Yet economic commentators and media always feel the need for some kind of overarching statistical measure of the economic performance of a nation. There is no end to the number of so-called "composite indicators" that have been proposed over the years, combining the strict GNP achievements with some more or less fancy measurement of the social achievements. In the pages to follow, we shall show how it is possible to extend the reach of the GDP *without* sacrificing its conventional definition. Rather than looking for new and ever more diffuse aggregates, we shall simply seek to document how well countries in the contemporary world are able to direct their productive capacity (as measured by their GDP) toward meeting the UN economic and social

* A **note to the reader**: Chapter 1 forms a self-contained essay that can be read independently, without consulting other chapters in the present volume. It discusses the motivation for forming an aggregate evaluation of the 17 United Nations sustainability goals (SDGs) pronounced by the *UN Agenda 2030* and describes the analytical framework. Using data for 28 EU nations, we calculate a measure of the effectiveness of policy in converting the GDP of a nation into the UN goals. The most successful nations in this respect (and they may include both rich and poor nations) provide benchmarks against which the efforts of other less successful ones can be compared.

Measuring Sustainable Development Goals Performance
ISBN: 978-0-323-90268-7
https://doi.org/10.1016/B978-0-323-90268-7.00002-5

SDGs. Our results are brought together in the calculation of an "effectiveness rating" of policy for each country.

The format of analysis proposed here hails back to the classical theory of economic policy as laid out in the early 1950s by Jan Tinbergen (1952 and 1956). In that spirit, we view GDP as an intermediary "means" influencing the SDGs ("targets"). Furthermore, following Tinbergen, we shall postulate a global social welfare function "the SDG welfare function" of the goals, to be maximized by the policy-maker.

Optimal policy, however, does not necessarily coincide with actual policy. In the real world, the actual goal achievements of a country often fall short of the optimal ones. To understand and capture this, we need to abandon Tinbergen's assumptions of general equilibrium, allowing deviations between optimal behavior and actual behavior. The economic and social policy of a country may be badly coordinated, or misdirected. It may be chaotic or even entirely without control (as during rebellion or war). The rankings will recognize equilibrium or disequilibrium behavior. Countries in disequilibrium will be ranked according to their relative success in reaching the optimal SDGs. Their shortcomings in reaching each SDG will be calculated.

Standard calculations of GDP provide no clue as to how a nation could possibly improve its performance. They meekly record actual achievements, whether in the upper range or the lower range. The great advantage of the effectiveness rating of policy to be defined here is that it is *diagnostic*: it holds specific suggestions as to how the sustainability achievements of a nation could be improved.

1.2 Doubting the blessings of economic growth

The idea that economic growth sows the seeds of its own destruction is an old one. It was a main theme of Thomas Malthus, who argued that improved living conditions in a region or a country will always be wiped out by a corresponding increase in population (*An Essay of the Principle of Population*, London 1798). Malthus actually formulated a mathematical law to this effect: the means of subsistence increases linearly but the population, if unchecked, increases exponentially.

In the contemporary world, many countries in Africa display the ailments that Malthus predicted, with a current total population of 1.2 billion people (the figure having doubled since 1990). The explosion in population was made possible by dramatic medical progress and in the cultivation of new and high-yield crops. Yet, there is widespread poverty and human misery. According to FAO, one third of the population in sub-Saharan countries is undernourished. The UN Population Fund writes: "In today's world, gaps in wealth have grown shockingly wide. Billions of people linger at the bottom, denied their human rights and prospects for a better life. At the top, resources and privileges accrue at explosive rates ..." (*Worlds Apart, State of the World Population, 2017,* New York, 2017)

The great German naturalist Alexander von Humboldt, traveling in America, saw another unfortunate consequence of economic growth: The rapid deforestation and the destruction of the forest's undergrowth dried the land and it became barren. The damage done by such human intervention was already "incalculable" (*Personal Narrative of Travels to the Equinoctial Regions of the New Continent*, London, 1814–1829). Humankind had the power to destroy the environment and the consequences could be catastrophic.

The Club of Rome report *The Limits to Growth*, 1972, presented the rapid economic growth the world was experiencing at the time as a key reason for the increase in global environmental problems such as pollution, shortage of raw materials, and the destruction of ecosystems. The same ideas were formalized by Georgescu-Roegen in his *The Entropy Law and the Economic Process* 1971, one of the founding works of the discipline of ecological economics. To avoid the rapid destruction of global resources, some of Georgescu-Roegen's followers advocated national policies of "Zero percent economic growth."

Today the world is confronted by yet another danger: global warming caused by burning fossil fuels, and the accumulation of harmful gases in the stratosphere. The Swedish Nobel laureate S. Arrhenius had already toward the end of the 19th century predicted the coming of global warming due to an increase of atmospheric carbon dioxide (the greenhouse effect). He actually derived the precise mathematical formula linking atmospheric carbon dioxide concentrations to the ground temperature of the globe (Arrhenius, 1896). Global warming is not "fake news". It is a scientific fact, known for a long time.

The Luddites in 18th century England experienced another negative side of growth: jobs lost in the textile industry due to the introduction of new weaving technology. Later generations of workers saw layoffs in the wake of the increasing mechanization of industrial jobs in the steel industry and in the manufacture of automobiles. More generally, technological change increases worker productivity and thus creates an incentive for management to reallocate (or lay off) workers. In the modern age, the advent of new communication and information technology has created new possibilities of distributing and organizing production on a global scale. Globalization has razed the city of Detroit, destroying some of the best paid industrial jobs in the world (the other side of the coin: the bustling IT industry in the city of Bangalore, the "silicon valley of India").

Some recent studies have argued that economic inequality has been on the increase in the developed world during the last 50 years, blaming the capitalist system itself (for example, see Pikketty, 2013). Technological innovation if successful bolsters the bottom line of farms, traders, and corporations. If the increased profit is returned to the owners, it will "trickle down" to consumers. But if it invested instead, worker productivity hopefully increases but wages may respond only gradually. Overall inequality in the economy increases, both in the distribution between the returns to capital and labor and in the vertical income distribution.

Present-day tech giants like Apple, Google, and Amazon pose other questions. The phenomenon of hyper-growth (the presence of vast economies of scale in internet-based enterprises) not only aggravates the trend toward inequality but also accelerates the flight of corporations to distant tax heavens. There is also the recent concern that the ever-growing data banks built by these giants undermine personal privacy.

The 2020 corona pandemic proves that serious health concerns of a nation can override even the incessant agenda of economic growth. That is indeed what the lockdown in the face of the pandemic signifies: both jobs and productive capacity are willingly sacrificed to protect lives.

As we shall argue in what follows, rather than blindly accepting GDP as the magic wand bringing progress, the world needs to learn to see GDP as a policy tool to be managed, aiming for global goals benefitting all.

1.3 The millennium development goals and the 2030 agenda for sustainable development

A series of international accords have brought attention to the wide variety of social and economic problems facing nations. A notable event occurred with the formulation of the UN Millennium Declaration, 2000, laying down a series of "millennium development goals" (MDGs).

The original eight MDGs were set for an intended implementation period of 15 years, from 2000 to 2015. The *Millennium Development Goals Report 2015* (UN, New York, 2015) reports on the main results. The publication *The Millennium Development Goals Indicator*, released January 2008 lists the indicators chosen to characterize each goal. It also contains a detailed data appendix.

The Rio + Conference held in June 2012 in Rio de Janeiro was an important milestone initiating a process of strengthening the institutional framework for policies aiming at sustainable development. The concluding document *The Future We Want* stressed the growing recognition that we live on a single planet within explicit planetary boundaries, and in an increasingly interdependent global multidimensional system that goes well beyond the economic marketplace. What future do we want? The UN answer: sustainability.

A collection of SDGs now replaces the MDGs. The final document *Transforming our World: the 2030 Agenda for Sustainable Development* was agreed on by world leaders in New York, September 2015. It lists 17 SDGs (shown here are the UN short hand designations):

SDG I no poverty
SDG II zero hunger
SDG III good health and well-being for people
SDG IV quality education
SDG V gender equality
SDG VI clean water and sanitation
SDG VII affordable and clean energy
SDG VIII decent work and economic growth
SDG IX innovation and infrastructure
SDG X reduced inequalities
SDG XI sustainable cities and communities
SDG XII responsible consumption and production
SDG XIII climate change
SDG XIV life below water
SDG XV terrestrial ecosystems
SDG XVI peace andjustice and strong institutions
SDG XVII partnership for the goals

The UN publication *2030 Agenda for Sustainable Development*, 2015 lists no less than 169 official indicators (called "targets") characterizing each SDG. Collecting all these data is a huge task, exceeding the possibilities of single research team. Relevant UN agencies are now involved in assembling this huge data repository for all countries in the world.

1.4 The SDG welfare function

Assuming that the SDG achievements of a nation have been measured, the next step would be to form some kind of index assessing its performance in comparison with other nations. The "index" that we are referring to here would be an omnibus representation of the total societal, environmental, and sustainability achievements of a country, a kind of modern version of what an older generation of economists used to call "social welfare" or "social utility." The great Dutch economist Jan Tinbergen, one of the fathers of economic statistics and econometrics, called for the coordination of available policy tools to maximize "social utility." Once having specified the social welfare function "as precisely as possible," he said, one should then "use it for finding the socioeconomic optimum" (quoted from his Nobel lecture, 1969).

In the spirit of Tinbergen, we now propose to aggregate the SDG data for a single nation into a representation of its "SDG welfare function." In a fashion, we shall actually be able to draw on some of the thinking about these things 100 years ago. Let us indeed assume that there exists a so-called social "utility function." Our great advantage is that the SDGs provide an internationally recognized list of the various dimensions of social utility. Denote the measured values of each goal or goal indicator by Y_r, $r = 1,2, \ldots,17$. The entire collection of goals is represented by $Y = (Y_1, Y_2, \ldots. Y_{17})$. Let us then construct the SDG welfare function of a nation as:

$$V = V(Y) \tag{1.1}$$

Returning to the idea of an index, one would in the first instance be tempted to form V as a linearly weighted average of the SDG observations. Conventional index calculations assume fixed and given weights, as for instance in the calculation of a cost-of-living index. Calculations of "social happiness" and similar welfare aggregates also use fixed weights of its various components, usually determined by "experts." Here, however, we encounter a problem: How do you rate the weight of basic nutrition (SDG II) compared to the weight of the right to free assembly (SDG XVI)? And why should these weights be the same for all nations, for the rich as well as the poor ones? An assumption of fixed SDG weights would be manifestly at variance with reality.

The SDGs are often substitutable at the margin in the sense that some shortcomings of one goal may be compensated by a more plentiful provision of one or several other goals. National goals emerge through a process of reconciliation between many competing points of view. A simple example of such substitutability is provided by the current political debate regarding the advisability of drilling for oil in arctic waters (in Alaska, Canada, Norway, and Russia). To what extent should drilling be limited to help preserve fish stocks and the pristine nature of the polar regions (SDG XIV), or expanded to provide employment and income opportunities for the local population (SDG VIII) ? The answer will always be a compromise, weighing the desirability of two societal goals against each other.

The marginal rate of substitution between two goals can be represented graphically as the slope of an indifference curve. Along each such indifference curve, the social preference $V =$ constant. The slope of the indifference curve at a given point indicates the rate of substitution between the SDGs. The curves are taken to be convex toward the origin, describing a negative substitution effect: as the coverage of one goal gets smaller, the policy maker needs to be compensated by a greater achievement in terms of one or more other goals.

An example of a welfare function that generates indifference curves that are convex toward the origin is the Cobb–Douglas function. For the calculations to follow, we shall assume that the function $V(Y)$ is generalize Cobb–Douglas, so that the relationship between social preference and the SDGs is linear in the logarithms of the variables:

$$\log V(Y) = \sum\nolimits_r \mu_r \log Y_r \tag{1.2}$$

where $\mu = (\mu_r, r = 1, 2, \ldots, 17)$ are weights to be determined empirically. Recording all data in the space of their natural logarithms, we thus end up calculating the SDG welfare of a nation as a weighted average of its individual SDG achievements.

1.5 Calculating the effectiveness of policy

The purpose of the calculations to be presented is to investigate the success (or lack of success) of nations in converting their GDP (here to be denoted X) into the sustainable development goals. We shall calculate the policy *effectiveness ratio* of each nation as the ratio between total SDG welfare achieved and the imputed economic and social policy cost. In simple words, the policy effectiveness of a nation is defined as the ratio between the total SDG welfare obtained, and the shadow cost of total GDP. Mathematically,

$$\sum\nolimits_r \mu_r \log Y_r / \nu \log X \tag{1.3}$$

where $\nu \log X$ is the shadow cost of the GDP. In simple words, the effectiveness of a nation is calculated as the ratio between the total SDG welfare obtained, and the shadow cost of total GDP.

The aim of economic and environmental policy is to make the ratio (Eq. 1.3) as large as possible—to maximize the SDG welfare function for any given level of social costs. The general idea here is that the ratio signals the relative success of policy: it is equal to one if all goals are being achieved, and less than one if one or several of these goal posts remain unfulfilled. A ratio greater than one would somehow indicate that the SDG welfare exceeded its imputed cost so that welfare could be obtained for free—which does not make sense:

$$\sum\nolimits_r \mu_r \log Y_r / \nu \log X \leq 1 \tag{1.4}$$

The mathematical details of the maximization problem facing each nation will be discussed in Chapter 2.

Rather than assuming the SDG weights to be fixed and known numbers (as in the calculation of a conventional index), we treat them as unknowns. The aim is to determine a unique set of goal weights μ_r for each nation r.

This optimization program reaches beyond the scope of the kind of economic and social policy that Tinbergen had in mind. Tinbergen's theory was essentially developed for one single country. He wanted to maximize the "utility function" of the Netherlands. In the same spirit, we maximize the policy effectiveness ratio for a single country, calculated as the ratio between its SDG welfare and its imputed GDP cost. The maximization is carried out subject to the condition that the ratio should be "meaningful"—a percentage number not exceeding 100%.

Now consider a list of nations, say $j = 1, 2, \ldots, n$. Denote their SDG achievements by Y_{rj}, $r = 1, 2, \ldots, 17$, $j = 1, 2, \ldots, n$. Their recorded levels of GDP (conventionally measured in US dollar equivalents) are X_j, $j = 1, 2, \ldots, n$. The nation currently to be rated will be identified as $j = 0$. Its ratio of policy effectiveness is $\Sigma_r \, \mu_r \log Y_{r0} / \, \nu \, \log X_0$. The maximization of this ratio has to be carried out subject to the condition that it be "meaningful"—a percentage number not exceeding 100%.

The mathematically interested reader is referred to Fig. 1.1 Employing the index weights determined for nation $j = 0$ and using the *very same* weights for every other nation j, the resulting effectiveness ratio should still remain meaningful.

The latter constraints tie all nations together. The SDG welfare function of a nation determined by the optimization program does not just reflect the goal preferences of that nation alone. The solution weights of the goals emerge as the result of an overarching optimization that includes information about every other nation, as a kind of *team consensus* about the priorities of the goals.

$$
\begin{aligned}
&Max\ \ \Sigma_r\,\mu_r \log Y_{r0}\ /\ v\log X_0\\
&\text{subject to}\\
&\Sigma_r\,\mu_r\ \log Y_{r0}/\ v\log X_0 \le 1\\
&\Sigma_r\,\mu_r\ \log Y_{rj}/\ v\log X_j \le 1\ ,\ j=1,2,\ldots,n\\
&\mu,\ v \ge 0
\end{aligned}
$$

Figure 1.1 The fractional programming problem. It took some of the smartest minds in mathematics to solve it. See Chapter 5.

Returning to Tinbergen's dictum in his Nobel lecture, we shall indeed "specify the social welfare function as precisely as possible," using it "for finding the socioeconomic optimum." The individual optimum for each and every country.

Unfortunately, however, Tinbergen may not have liked our mathematical results. Yes, a fair number of countries turn out to achieve the maximal policy effectiveness rate of one. These countries may display *Pareto optimality,* a state of classical equilibrium named after the Italian economist Vilfredo Pareto (1848—1923). In the present case, it would refer to a situation where a country reaches a maximal rate of policy effectiveness equal to one and no actual goal achievement must fall short of the optimal goal. No goal slack is permitted.

Tinbergen's conception of the world was different from ours: an orderly world of good will and equilibrium, a rosy world where every country would be governed by a well-meaning planning board seeking to maximize the welfare of its inhabitants. An optimal world where the rate of effectiveness of policy would always equal one and every nation would be Pareto efficient.

Our statistical results tell another story. As we shall see, a fair number of nations end up with a maximal policy effectiveness ratio less than one, and/or goal slack. They are not Pareto optimal. They present instances of disequilibrium rather than equilibrium.

1.6 A brief numerical look

For the numerical work to be reported here, we used the most recent data on the 28 EU nations (the EU in 2019 still comprising the United Kingdom) assembled by Sustainable Development Solutions Network and Institute for European Environmental Policy See European Commission,

Eurostat *Sustainable development in the European Union—Monitoring report on progress toward the SDGS in an EU context*—2020 edition.

With data for only 28 nations, it is necessary to limit the number of SDG indicators. We decided to pick as diverse SDGs as possible, representing both human life characteristics, social conditions, and climate and environmental concerns.

Table 1.1 lists GDP and six SDG indicators, for each EU nation as follows:

- ***GDP***—gross domestic product per capita, in 2017 US dollar equivalents (purchasing power parity).
- ***LIFE***—an indicator of SDG VIII, life expectancy at birth, in years (OECD code name "sdg3_lifee"), latest data for 2017.
- ***NEET***—an indicator of SDG VIII, Youth not in employment, education or training, % of population aged 15−24, per cent of total labor force (OECD code name "sdg8_euyneet"), latest data for 2018.
- ***GINI***—an indicator of SDG X, being the conventional GINI coefficient adjusted for top income ("sdg10_adjgini"), latest data 2014. The Gini coefficient reflects the degree of inequality of the vertical income distribution in a country.

Table 1.1 Actual practice = best practice for 17 Pareto optimal EU nations. SDG10 and SDG13 measured in their original unadjusted scales. Only one decimal shown here.

	GDP	LIFE	NEET	GINI	LEEK	EMITCO2	BIODIV
Bulgaria	18,606	74.9	18.1	41.1	13.0	5.8	98.9
Croatia	22,828	78.0	15.6	38.2	11.2	4.0	74.1
Cyprus	22,828	78.0	14.9	38.2	29.3	4.0	66.1
Czechia	32,415	82.2	9.5	35.5	7.7	5.4	92.3
Denmark	46,683	81.1	8.5	28.8	16.4	5.8	89.7
Finland	40,586	81.7	10.1	28.3	4.6	8.7	74.8
France	38,606	82.7	13.6	32.6	11.1	5.0	80.9
Greece	24,574	81.4	19.5	44.9	12.9	5.4	85.8
Italy	35,220	83.1	23.4	38.2	16.1	5.5	77.9
Lithuania	29,524	75.8	9.3	45.4	15.7	4.5	90.5
Latvia	25,054	74.9	11.6	35.9	23.5	3.5	97.3
Malta	36,505	82.4	7.4	29.4	7.1	6.3	99.3
Netherlands	48,473	81.8	5.7	29.4	15.7	10.4	90.6
Portugal	27,937	81.6	9.6	42.6	26.9	4.6	74.1
Romania	23,502	75.3	17.0	52.4	10.1	3.4	77.3
Slovakia	30,155	77.3	14.6	33.4	6.7	5.5	82.7
Slovenia	31,401	81.2	8.8	27.5	22.7	6.4	85.1
Spain	34,272	83.4	15.3	38.5	15.9	5.1	56.6
Sweden	46,949	82.5	7.0	27.5	7.8	4.4	58.4

- ***LEEK***—an indicator or SDG XI, Population living in a dwelling with a leaking roof, damp walls or foundation, or rot in window frames or floor, percentage of total population ("sdg11_dwelling"), latest data for 2018.
- ***EMITCO2***—an indicator of SDG XIII, measuring the amount of air pollution, tons of CO_2 emitted per capita, ("sdg13_co2pc"), latest data for 2017.
- ***BIODIV***—an indicator of SDG XV, measuring mean area that is protected in terrestrial sites important to biodiversity, per cent ("sdg15_cpta"), latest data for 2018. The data base features three indicators of biodiversity, relating to marine sites, terrestrial sites, and freshwater sites; we arbitrarily picked one of them.

The SDG indicators chosen here reflect the strange habit of the UN statisticians of mixing desirable goals with undesirable ones. But for programming purposes we needed to convert LIFE, NEET, GINI, LEEK, EMITCO2, and BIODIV to something wanted rather than unwanted. This is easy when the indicators are simple percentage points, as for NEET, GINI, LEEK. Using the number 100 - GINI rather than the Gini coefficient itself will reflect the degree of equality in the vertical income distribution rather than its dispersion. As for the emission of CO_2, the indicator EMITCO2 measures the CO_2 "footprint" per capita released in a nation. Again, we had to convert the official figures to something desirable, this time representing the degree of absence of CO_2 in the air. Hence, we formed 20 EMITCO2, picking the number 20 tons as an admittedly arbitrary upper limit on the emission of CO_2. Altogether then, we formed the desirables LIFE, 100 − NEET, 100 − GINI, 100 − LEEK, 20 − EMITCO2, and BIODIV.

The mathematics of the multidimensional methods employed in the present book has already been briefly characterized, culminating in the text box Fig. 1.1. One additional observation should be made. These methods allow for two alternative index calculations to be made: one assuming constant returns to scale (in the present case meaning that a doubling of GDP should also lead to a doubling of all SDG achievements), the other assuming only variable returns to scale. For the present, we decided to choose the latter alternative, permitting a closer fit when the number of observations is limited. See Chapter 2.

Writing computer code using the commercial software package General Algebraic Modeling System, and maximizing the policy effectiveness ratio for the 28 EU countries, we obtained numerical results reported below. Nineteen countries reached a policy effectiveness ratio equal to one:

Bulgaria, Croatia, Cyprus, Czechia, Denmark, Finland, France, Greece, Italy, Lithuania, Latvia, Malta, Netherlands, Portugal, Romania, Slovakia, Slovenia, Spain, Sweden (1.5)

They are all Pareto optimal, displaying neither any GDP slack nor any goal slack. They are all listed in Table 1.1. For each of them, the recorded GDP and the recorded SDG data coincide with the optimal GDP and the optimal SDG achievements as calculated by the optimization model.

Let us take a look at two of those countries, Bulgaria and Romania. They both report life expectancies less than 75 years, 17% or more of the youngsters excluded from the labor market, a high degree of inequality in the income distribution, average results for the quality of living quarters, good marks for the CO_2 footprint (less than six tons per capita), and average (Romania) to outstanding (Bulgaria) marks for terrestrial biodiversity. Altogether not very impressive results, and yet these two nations are ranked among the winners.

The simple reason is that Bulgaria and Romania together with 17 others won the race *of converting their GDP to SDG achievements*. Since those two nations together with Cyprus have the lowest GDPs of all EU nations (less than 24,000 US dollar equivalents per capita), it takes only moderate SDG achievements to score an excellent conversion ratio.

With increasing GDP, the conversion race becomes more competitive. Denmark, Finland, the Netherlands, and Sweden all have GDPs in the 40,000–50,000 dollar brackets. But they also scored some impressive SDG results, like Denmark, and Sweden being among the most equitable EU countries, Finland having the best made and maintained private homes, and the Netherlands by far being the most successful EU country in keeping its youngsters employed, in education or in training (NEET).

Croatia, Cyprus, Latvia, and Romania obtained the most favorable EU results in terms of emission of CO_2 (less than four ton of CO_2 per capita), which certainly helped to bring these nations into the Pareto elite. Their GDPs per capita are all low, demonstrating that a moderate GDP often is a friend of the environment.

The longest life expectancy among all the 28 EU nations is reported in Italy and Spain (more than 83 years), the shortest in Bulgaria, Lithuania, and Romania. The most egalitarian countries are Denmark, Finland, Slovenia, and Sweden. Finally, the lowest CO_2 footprints belong to Croatia, Latvia, and Romania, with Estonia and Luxembourg far behind, spewing out more CO_2 than the others.

Taken together, the nations (Eq. 1.5) define a *frontier* between the optimal and the suboptimal. Any weighted linear average of those nations with the weights summing to one represents a hypothetical composite combination that would also be optimal. But if the weights sum to less than one, that hypothetical country falls *behind the frontier* and would be suboptimal.

Let us take a look at two of the frontier countries, Bulgaria and Romania. They both report life expectancies less than 75 years, 17% or more of the youngsters excluded from the labor market, a high degree of inequality in the income distribution, average results for the quality of living quarters, good marks for the CO_2 footprint (less than six tons per capita), and average (Romania) to outstanding (Bulgaria) marks for terrestrial biodiversity. Altogether not very impressive results, and yet these two nations are ranked among the winners.

With increasing GDP, the conversion race becomes more competitive. Denmark, Finland, the Netherlands, and Sweden all have GDPs in the 40,000–50,000 dollar brackets. But they also scored some impressive SDG results, like Denmark, Malta, Slovenia, and Sweden being the most equitable EU countries, Finland having the best made and maintained private homes, and the Netherlands by far being the most successful EU country in keeping its youngsters employed, in education or in training (NEET).

1.7 The frontier

Inspecting the list (Eq. 1.5) further, the idea of comparing one of those Pareto-optimal countries with another lies near at hand. For any given pair, one may form a simple linear average, giving each of them the weight of 50%. For instance, comparing the data for Bulgaria and Cyprus, one might form the average 0.5 × Bulgaria + 0.5 × Cyprus. The result is data for a hypothetical country that does not exist, of course. But it is always tempting to compare one country to another, to see if one of them possibly has something to learn from the performance of the other. Such a weighted average would not represent any actual practice but a hypothetical practice that would be achievable or "doable" in the sense that it would be a straightforward linear combination of the observed practice of real countries. The general mathematical form of such linear interpolations would be

$$\text{weight} \times \text{Bulgaria} + \text{weight} \times \text{Croatia} + \text{weight} \times \text{Cyprus} + \text{weight} \times \text{Slovenia} + \text{weight} \times \text{Spain} + \text{weight} \times \text{Sweden} \tag{1.6}$$

with the weights summing to one. Each hybrid (Eq. 1.6) is a case of "best practice." Examining such hypothetical combinations of SDGs is not just an intellectual game but actually the essence of policy planning: searching for alternative optimal policies that would yield maximal SDG ratings.

Geometrically, the situation may be explained as follows. We are given data for 28 EU countries. For each country, we have observations of one input (the measured GDP of the country) and of six outputs (the SDG indicators). So, each observation is a point in a seven-dimensional space. Any linear average (Eq. 1.6) is located on some *linear facet* connecting the 28 observed points. The totality of all linear facets forms a kind of *frontier* or border-line between achievable goal combinations and nonachievable ones.

The frontier is a multifaceted multidimensional surface, not unlike the visual organ of a fly which is comprised of thousands of closely packed hexagons on the surface of the eye. The surface consists of linear facets and corner points. The list of 19 nations (Eq. 1.5) constitute the corner points. They span the domain of achievable goal combinations.

1.8 Subfrontier nations

The nine countries listed in (Eq. 1.7) below obtained a policy effectiveness ratio less than one. They did not manage to translate their GDP into optimal SDG ratings.

Austria, Belgium, Germany, Estonia, Hungary, Ireland, Luxembourg, Poland, UK (1.7)

They are Pareto nonoptimal, in disequilibrium. Geometrically, they are located below the frontier. Their effectiveness ratios are listed below:

Ireland, Luxembourg, Poland 0.999

Belgium 0.998

Germany, Estonia 0.997

Austria, Hungary, UK 0.996

For instance, when the UK obtains the score 0.996, this means that its equilibrium SDG values would have been 1.004 times greater.

The results now reported are surprising. The suboptimal countries include some of the most modern Western countries like Austria, Belgium, Germany, Ireland, and Luxembourg, all with a GDP exceeding 40,000 dollar equivalents per capita.

The case of the UK being at the bottom of the list is another surprise. Examining Table 1.1 we see that the UK scores fairly average results for

Table 1.2 Recorded data (the upper number in each cell) and calculated best practice (the lower number) for nine suboptimal EU countries. All SDG numbers measured in their original scales, rounded to one decimal.

Country	GDP	LIFE	NEET	GINI	LEEK	EMITCO2	BIODIV
Austria	45,437	81	8.4	32.0	10.4	6.8	66.6
	45,365	82.4	6.9	28.1	8.9	5.8	67.7
Belgium	42,659	81.6	12	29.8	18	8.6	81
	39,544	82.5	7.6	29.1	7.6	5.7	81.9
Germany	45,229	81.1	7.9	33.4	12.5	8.8	78.3
	44,500	82.2	6.6	28.7	10.7	7.5	79.4
Estonia	29,481	78.4	11.7	37.4	13.6	14.2	94.8
	29,481	79.5	10.5	34.9	12.4	5.7	96.1
UK	39,753	81.3	11.7	37.7	17	5.8	84.3
	37,806	82.6	10.3	31.1	9.2	5.6	85.6
Hungary	26,778	76	12.9 11.1	36.2	22.5	4.7	83.1
	26,778	77.5		35.0	21.0	4.5	91.3
Ireland	67,335	82.2	11.6	32.9	12.6	7.8	87.7
	35,978	82.6	11.1	32.1	9.9	6.0	88.2
Luxembourg	94,278	82.1	7.5	31.7	17.4	15.1	83.3
	40,241	82.4	7.2	28.8	7.9	6.1	83.6
Poland	27,343	77.8	12.1	43.9	11.6	7.6	87.6
	27,343	78.1	11.8	37.6	11.3	5.7	97.0

every one of the six SDGs, neither any striking failure nor any striking success. But just being average did not suffice to bring the UK into the Pareto elite.

In order to discuss nature of the results now reported, we need to introduce the idea of "projecting" a suboptimal country onto the frontier. The projection is carried out maintaining the recorded GDP of the country but increasing the SDG achievements equiproportionally until it hits the frontier. Conventionally, the projection point is referred to as "best practice" for the country considered, just as every single point on the frontier is referred to as "best practice." But remember that it is a hypothetical point, a sort of "composite country" of the form (Eq. 1.6) above.

The projected frontier values of each suboptimal country (Eq. 1.7) are listed in Table 1.2 below.

1.9 Interpretation

Standing back for a moment now and considering the results of the multidimensional analysis, the following question immediately presents itself: Why is it that an expanding GDP in modern Western countries like Belgium, Germany, and the UK does not suffice to reach optimal sustainability goals?

The obvious answer is that GDP can be many different things, and that some forms of GDP creation are more conducive to meeting the UN goals than others. GDP arises as the sum of all value added created by economic production in a country, in manufacturing, agriculture, energy creation, transport, tourism, financial services, and so on.

The case of financial services is an eye-opener. The economies of Ireland and Luxembourg are crucially dependent upon their financial industry. Financial services may have many attractive features, but clearly they do not include furthering the sustainability accomplishment of a country. Actually, a burgeoning workforce in the financial sector may crowd out the recruitment to health care, care for the elderly, and other sustainable occupations. (As shown in Table 1.2, equilibrium GDP in Ireland and Luxembourg would be about a half of the recorded GDP. The other half of GDP does not translate into any gains of SDG. And yet these two countries are both badly underperforming in terms of the goal LEEK (private housing poorly insulated and/or deteriorating).

The energy sector is a major contributor to the GDP of a nation. But several EU nations are at loggerheads how to generate the energy they need while limiting the negative effects on the sustainability goals. The case of Poland is instructive, being heavily dependent upon coal as an energy source. While several EU countries are committed to phase out the use of coal within two decades (such as Germany), Poland generates around 80% of its power from coal (See the EU report *Overview of National Coal Phase-Out*, published on the website beyond-coal.eu, February 2020).

Or, consider tourism, so important to EU countries like France, Greece, Italy, Portugal, and Spain. They all compete to bring in ever greater numbers of free-spending tourists. Tourism surely is a great cash cow, but which are the sustainability consequences? Take Portugal, with clean beaches, pristine waters and with a proud national history documented in outstanding museums and still standing medieval cities. National hotel chains are engaged in a construction boom. And then there is golf tourism and health tourism. Rapidly proliferating golf courses in Portugal are the equal to any in the world. They need lots of irrigation, and the water table is sinking. The water bills paid by the ordinary Portuguese are increasing. No sustainability there.

Health tourism to Portugal is becoming more important for every year. Large, private hospital chains are building modern shining new facilities with all the high tech equipment catering for English, Dutch, and German short-term visitors. Sizable contributions to the measured GDP for sure, but helping little to improve the health of the Portuguese population who belong to the overcrowded state hospital sector.

Unfortunately, there is no statistical evidence on these matters and so we are left with case stories and anecdotal evidence. Hopefully, other researchers will conduct in-depth multidimensional analysis of the sustainability consequences of value added creation in individual sectors.

1.10 Concluding thoughts

As we are writing these lines, the COVID-19 virus is ripping through the social fabric of the countries of the world. It leaves bare horrifying stories of poverty, undernourishment, and lack of medical and social care even in the wealthiest of nations.

The calculations reported here show that GDP-rich EU nations like Austria, Belgium, Denmark, Germany, Ireland, and Luxembourg harbor pockets with considerable SDG achievement deficits.

Perhaps this means that economic and social policy all over the world has been misdirected. Established economic theory has failed to see the new social realities of the 21st millennium. Macroeconomic theory founded by J.M. Keynes in the 1930s considers GDP and national income as the key economic variables driving household and government consumption, capital formation, exports, and economic progress. In terms of conventional economic theory, the economic hardship of the pandemic will be seen as just an (albeit serious) cyclical downturn.

Obviously a growing GDP is no panacea for social, environmental, and educational success. Mainstream economists seem to have shut their eyes to the opening of a widening abyss of poverty, malnourishment, and rejection amidst our supposedly "rich" societies. Cities are riddled with violent crime, drug running, and human trafficking. National economies have segregated into ethnic pockets of alternative cultures with endemic poverty continuing from one generation to the next, surviving on faltering social protection systems.

Our empirical work reported here and elsewhere demonstrates the weak links, if any, between economic growth and the true indicators of national economic welfare, as spelled out by the UN sustainable development goals. As the world plunges into its deepest GDP recession ever recorded, an opportunity presents itself to redirect the priorities of the economic system, returning to basic aims of human well-being and social compassion.

In a series of publications, the Nobel laureate and former adviser to President Clinton, Joseph E. Stiglitz has argued that economists need new metrics that go beyond the single measure of GDP:

> *It is clear that something is fundamentally wrong with the way we assess economic performance and social progress…If our measures tell us everything is fine when it really isn't, we won't make the right decisions. And it should be clear that, in spite of the increases in GDP… everything is not fine.*
>
> *…we argued that economists and governments could not capture anything as complex as our society in a single number. We needed a dashboard … to reflect the many dimensions of success or deprivation — including inequality, economic insecurity and sustainability.*
>
> ***quoted from his preface to Stiglitz, J. E., Fitoussi, J.-P. & Durand, M. (2019).* Measuring What Counts, The Global Movement for Well-Being. *New York: The New Press.***

This does not mean that GDP has to be dismissed as a key variable in the economic system. GDP is not only the fruition of past and current efforts, but also as the germ of new progress. The future lies with constructive GDP *policy*. Nations need to develop a new kind of GDP management.

In the text presented here, we have indicated the nature of such policy. We have shown how it is possible to calculate the policy effectiveness ratio of each nation as a measure its relative success in converting its GDP to the UN Sustainable development goals.

CHAPTER TWO

Diagnostics for economic and social policy

2.1 International initiatives

A series of international accords has recently brought attention to the wide variety of social and economic problems facing nations in a world of globalization, economic development (or lack of development), and scarce resources. The need for policy coherence, coordination, and cooperation has intensified during past years. A notable event occurred with the formulation of the United Nations Millennium Declaration (New York) (2000). It laid down a series of "millennium development goals" (MDGs), giving a sharper focus and impetus to the international quest for social and economic development.

The Millennium Declaration resolved "to ensure greater policy coherence and better cooperation between the United Nations (UN), its agencies, the Bretton Woods Institutions and the World Trade Organization, as well as other multilateral bodies, with a view to achieving a fully coordinated approach to the problems of peace and development." The development community reached a consensus around the shared objectives of achieving the MDGs and the partnership between developed and developing countries:

- the developing countries are expected to focus on good governance, sound economic and social policies, and robust institutions that will help mobilize resources from all sources—public, private, foreign, and domestic;
- the developed countries are expected to reduce barriers to trade and promote an international economy more conducive to growth and development, to provide adequate aid resources in the form of private capital flows, to provide additional development assistance to support those efforts, and (under the Heavily Indebted Poor Country Initiative) help the developing world to ease the debt burdens and support them in effective management of their external debts (see also UN Department of Economic and Social Affairs (UN, 2004)).

Measuring Sustainable Development Goals Performance
ISBN: 978-0-323-90268-7
https://doi.org/10.1016/B978-0-323-90268-7.00017-7

The MDGs include poverty reduction, full and productive employment and decent work for all, educational goals, health goals, and other measures. The Monterrey Consensus of the UN (2004) International Conference on Financing for Development emphasized the need for coherent action of all stakeholders in the international community, including the need to enhance the efficacy of macroeconomic policies. The Millennium Declaration was followed by a report by Secretary-General Kofi Annan titled, *Road Map toward the Implementation of the United Nations Millennium Declaration* (UN, 2001b), which brought together the goals listed in the declaration under eight separate chapters:

- MDG 1: Eradicate extreme poverty and hunger;
- MDG 2: Achieve universal primary education;
- MDG 3: Promote gender equality and empower women;
- MDG 4: Reduce child mortality;
- MDG 5: Improve maternal health;
- MDG 6: Combat HIV/AIDS, malaria, and other diseases;
- MDG 7: Ensure environmental sustainability;
- MDG 8: Develop a global partnership for development.

Founded in Geneva in 1919, the International Labour Organization (ILO) is a specialized agency of the UN to promote opportunities for women and men to obtain decent and productive work in conditions of freedom, equity, security, and human dignity. Working through dialog with governments, employers, and workers, the ILO has set international labor standards that form part of international human rights law and national legal systems. From its inception, through the means of research, advocacy, and technical cooperation, the ILO seeks the promotion of social justice and internationally recognized human and labor rights. The ILO constitution, *Declaration of Philadelphia* [2004], states:

> *… the attainment of the conditions in which … [social justice] shall be possible must constitute the central aim of national and international policy … it is the responsibility of the International Labour Organization to examine and consider all international economic and financial policies and measures in the light of this … fundamental objective.*

In his 1999 annual report, the director-general of the ILO, Juan Somavia, set out the *Decent Work Agenda* (ILO, 1999) with four elements:

(1) *employment*: the principal route out of poverty is through work and income;

(2) *rights*: without them, people will not be empowered to escape from poverty;

(3) *social protection*: it safeguards income and underpins health;
(4) *dialog*: the participation of employers' and workers' organizations.

Since then, the concept of "Decent Work" has provided a strategic framework for organizing ILO's activities as the simultaneous pursuit of these four objectives. The concept of Decent Work draws on the long experience that the ILO has gained, working with constituents in a variety of programs and in interaction with employment, labor markets, and social protection policies in many countries and under many different circumstances (ILO, 2001a).

The logical relationship between the MDGs and various instruments of economic and social policy may be illustrated by a cognitive map as in Fig. 2.1. The cognitive map is a common tool in multicriteria analysis (see e.g., Bana e Costa and Thore (2002)). Figure 2.1 highlights the logical structure of the causal relationships, tying causes to effects.

For 15 years, the MDGs drove progress in several important areas: reducing income poverty, driving down child mortality and improving maternal health, lowering the incidence of malaria and tuberculosis. However, the record also showed shortcomings of global development such as the increase of global emissions of CO_2 by over 50%. Several indicators pointed at an increasing distributional inequality, including the share of poorest quintile in national consumption and the growth rate of GDP per person employed. In short, many regions did register improvements in wealth and living standards, but the environment and distributional justice were set on a degrading path.

This uneven outcome was clear years before the end of the MDGs. In June 2012 participants at the Rio+20 conference in Rio de Janeiro decided to replace the MDGs with a set of sustainable development goals (SDGs), thereby returning sustainability considerations to the center of development thinking.

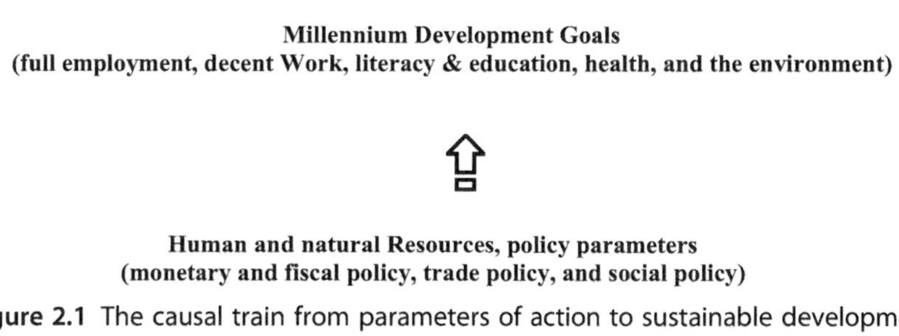

Figure 2.1 The causal train from parameters of action to sustainable development.

UN Secretary General Ban-Ki Moon launched the UN Sustainable Development Solutions Network, designed to mobilize a broad range of global scientific and technological expertise. In January 2016, the SDGs officially came into force. The SDGs were seen as an urgent call to shift the world onto a more sustainable path. The objective is to produce a set of universal goals that meet the urgent environmental, political, and economic challenges facing our world.

Compared to the MDGs, the SDGs are not only broader and far more committed to connecting economic improvements with their environmental and socioeconomic context. The new goals include inequalities, economic growth, decent jobs, cities and human settlements, industrialization, oceans, ecosystems, energy, climate change, sustainable consumption and production, peace, and justice. Many of the new goals address conditions where current trends are negative, such as climate change, global arms expenditures, deforestation, desertification, waste production, or road traffic deaths.

We reproduce below the list of SDGs as presented by the UN Statistical Commission in their 2019 Progress Report together with a limited number of statistical indicators that provide an overview of global and regional trends.

- SDG 1: End poverty in all its forms everywhere,
  - eradicate extreme poverty for all people everywhere,
  - achieve substantial social protection coverage.
- SDG 2: End hunger, achieve food security and improved nutrition, and promote sustainable agriculture,
  - end hunger,
  - end stunting in children under five years of age,
  - increase government investment in agriculture.
- SDG 3: Ensure healthy lives and promote well-being for all at all ages,
  - increase the coverage of births attended by skilled health personnel,
  - end preventable deaths of children under five years of age,
  - end the epidemic of HIV infections,
  - end the epidemic of malaria,
  - increase vaccine coverage of target population.
- SDG 4: Ensure inclusive and equitable quality education and promote lifelong opportunities for all,
  - ensure that all children achieve proficiency in reading at the end of primary education,
  - ensure that all children achieve proficiency in mathematics at the end of primary education,
  - reduce the out-of-school rate for primary school children.

- SDG 5: Achieve gender equality and empower all women and girls,
  - eliminate child marriage,
  - ensure women's full participation and equal opportunities in the labor force.
- SDG 6: Ensure availability and sustainable management of water and sanitation for all,
  - achieve universal and equitable access to safely managed drinking water services,
  - achieve universal and equitable access to safely managed sanitation services,
  - ensure the sustainability of water use by decreasing the level of freshwater withdrawal.
- SDG 7: Ensure access to affordable, reliable, sustainable and modern energy for all,
  - achieve universal access to electricity,
  - increase the share of renewable energy in the final energy consumption, double the global rate of improvement in energy efficiency.
- SDG 8: Promote sustained, inclusive and sustainable economic growth, full and productive employment and decent work for all,
  - sustain per capita economic growth,
  - achieve full and productive employment for all,
  - substantially reduce the proportion of youth not in employment, education, or training.
- SDG 9: Build resilient infrastructure, promote inclusive and sustainable industrialization, and foster innovation,
  - significantly raise industry's share of GDP,
  - reduce CO_2 emissions per unit of GDP,
  - substantially increase the expenditure for scientific research and development.
- SDG 10: Reduce inequality within and among countries,
  - reduce inequality within countries.
- SDG 11: Make cities and human settlements inclusive, safe, resilient, and sustainable,
  - reduce the proportion of urban population living in slums.
- SDG 12: Ensure sustainable consumption and production patterns,
  - reduce the domestic material consumption per unit of GDP ratio.
- SDG 13: Take urgent action to combat climate change and its impacts,
  - reduce global greenhouse gas emission.

- SDG 14: Conserve and sustainably use the oceans, seas, and marine resources for sustainable development,
  restore the proportion of fish stocks within biologically sustainable levels,
  increase coverage of marine protected areas.
- SDG 15: Protect, restore, and promote sustainable use of terrestrial ecosystems, sustainably manage forests, combat desertification, and halt and reserve land degradation and halt biodiversity loss,
  ensure the conservation, restoration, and sustainable use of forest areas,
  protect and prevent the extinction of threatened species.
- SDG 16: Promote peaceful and inclusive societies for sustainable development, provide access to justice for all and build effective, accountable and inclusive institutions at all levels,
  significantly reduce all forms of violence, including homicides,
  reduce the proportion of unsentenced detainees,
  increase the proportion of countries with independent national human rights institutions.
- SDG 17: Strengthen the means of implementation and revitalize the global partnership for sustainable development,
  ensure implementation of the net official development assistance commitments by donor countries,
  enhance access to technology by increasing fixed internet broadband subscriptions,
  increase proportion of countries with a national statistical plan.

Notice that the indicators do in no instance exhaust the possible coverage of the goal. Many more indicators could easily be mentioned (actually, the Statistical Commission defines no less than 169 "targets" characterizing the SDGs). In its annual progress report, however, the Commission limits itself to a few ordinal measures of each SDG such as "high progress," "moderate progress or "low progress."

2.2 Tinbergen on economic policy

How can a nation use economic and social policies to reach for the SDGs? In order to answer that question, it is helpful to recall the inherited classical theory of economic and social policy, formulated in the early 1950s by one of the giants of economic theory—the Dutch economist, Jan Tinbergen.

In the early 1930s, the League of Nations engaged Tinbergen to evaluate empirically and test the proliferating flora of business cycle theory of the times. It is easy to understand how this work gradually led him to develop a theory of economic planning. Behind the work on business cycles was the constantly nagging question of how it might be possible by the central bank or the government to counteract the onset of apparently ever-deepening business downturns. His models were large systems with as many as 50 relations and 70 variables; these included many variables that could be considered as "instruments" of policy, like financial aggregates, interest rates, and data on the government budget. Could such variables be manipulated by economic policy in order to dampen business fluctuations?

Appointed after the war as director of the newly established Central Planning Bureau for the Netherlands, Tinbergen was expected to forecast the effects of economic policy and to advise the government on related matters. It led to the writing of two key tracts on the theory of economic policy (Tinbergen, 1952, 1956). The 1952 book is the slimmer of the two, only 78 pages. He begins by explaining the nature of the model to be used. It contains four types of quantified entities: targets (goals), instruments, other variables of the economic system, and data. He goes on to argue that consistent economic policy requires that the number of targets equal the number of instruments. If the number of targets exceeds the number of instruments, the targets are "incompatible." If the number of targets is smaller than the number of instruments, the instruments are "alternative," that is, one instrument may be used instead of another or a combination of others. (We should point out that these assertions do not hold up in the nonoptimal framework that we will describe in subsequent chapters.) The book used simple examples illustrated with the aid of two models, one of which was used by the Central Planning Bureau.

The 1956 work introduces welfare functions for policymakers or entire nations. For the simplest case, when the number of targets equals the number of instruments, the mathematics goes like this (using Tinbergen's own notation). Let:

x_i, $i = 1, 2, \ldots, I$ be the economic variables that are not targets

y_j, $j = 1, 2, \ldots, J$ be the targets where $I = J$

z_k, $k = 1, 2, \ldots, K$ be the instruments controlled by a policymaker

u_l, $l = 1, 2, \ldots, L$ be the data.and write the equations of the model:

$$\Phi_n(x, y, z, u) = 0, n = 1, 2, \ldots, N \tag{2.1}$$

where x, y, z, u are vectors as defined a moment ago. If the model is "consistent," the number of equations equals the number of unknowns, $I + K = N$. Under standard regularity conditions, the model will then possess a well-defined and unique solution. The general format is illustrated by a great number of examples, ranging from monetary and currency policy to tax policy. In all cases, the problem is considered not only theoretically but also numerically, using plausible values of the coefficients derived from econometric work.

When the number of targets exceeds the number of instruments, however, solution requires the presence of a social welfare function, say:

$$\theta(y, z) \tag{2.2}$$

The problem at hand, then, is one of maximizing θ (y, z) subject to the side conditions (Eq. 2.2). In almost all applications, however, he assumes fixed targets so that there is little practical maximization in the book.

Tinbergen's work was an immediate sensation. The prestigious *Journal of the American Statistical Association* asked one of the young and coming economists at Stanford by the name of Ken Arrow to write a detailed report on the book (Arrow, 1958). One future Nobelist reviewing another future Nobelist! In many ways, the present reader may find the referee report as thought-provoking as the original work. Arrow starts out by examining Tinbergen's discussion of the aims or goals of economic policy, listing the following goals:

- international peace;
- maximum real expenditure per capita with full employment and monetary equilibrium;
- emancipation of underprivileged groups;
- "as much personal freedom as compatible with the other aims."

Agreeing that peace should be treated as an overarching aim, Arrow nevertheless finds it difficult to see how the economics profession can contribute to it. (With 60 years of hindsight, we can only admire the cleverness of the authors of the SDGs who were able to articulate a series of economic and social goals that all directly and without question contribute to peace.)

Discussing the very last point in the list—about personal freedom—Arrow makes a very poignant observation: the freedom of one person may be seen as compulsion by another:

> *I think the matter is more complex than is sometimes supposed. Man is a social animal and the freedom in question must extend to organizations of various kinds, not excluding labor unions.*
>
> ***[ibid., p. 91]***

In other words, some policy goals may be inherently contradictory. We shall later return to this issue in some detail, discussing the goal of free trade. (We shall also point at possibilities of resolving such conflicts of aims, breaking up the causal link from policy instruments to policy aims into two or several sublinks.)

Arrow reviews in some detail Tinbergen's discussion of "qualitative economic policy," where the institutions and methods of economic control are altered. This includes built-in stabilizers and changes in the types of taxation, monopoly policy, administrative centralization or decentralization, and appraisal of investment projects. Widening the scope of his investigation even further, Tinbergen also considered social security schemes, guarantees of minimum income or minimum employment, equalization of opportunities, monetary reforms, centralization of production decisions, industrial democracy, and nationalization. To the modern reader, the issue of qualitative economic variables boils down to the problem of identifying bundles of suitable quantitative target variables characterizing the situation at hand. Admittedly, this may involve a very large number of variables interacting with each other through complex interdependent causal networks. The cognitive map in Fig. 2.1 is then far too simplistic and needs to be replaced by a detailed mapping of the causal structure.

During the 1950s and 1960s, Tinbergen's international fame was growing and his work on economic planning won general acclaim. It was only natural that he (together with Ragnar Frisch) would be selected in December 1969 as the first recipient of the newly created Prize in Economic Science in memory of Alfred Nobel, instituted by Sveriges Riksbank (the central bank of Sweden) to celebrate its 300th anniversary.

Tinbergen's Nobel lecture, *The Use of Models: Experience and Prospects* (1969), must have surprised the listeners, as they heard him stress the limited reach of economic knowledge. Rather disarmingly, he confessed that his decade-long statistical studies of the fluctuations in investment activity during the business cycle never led to any confident conclusions and that, as head of the Netherlands central planning bureau, he found it safer to "ask industrialists for their investment programs" rather than to rely on econometric explanations. He had also found government expenditure difficult to forecast. (Both variables are determined by a small number of decision makers.) Random fluctuations are important.

To Tinbergen, social policy was a major determinant of the development of the economic system. Discussing the need to expand the economic analysis to include educational variables, Tinbergen pointed out that the budget for education in the Netherlands at his time was among the largest of the various ministries. *Educational variables* to be included in the analysis could be gross enrollment data in schools or educational facilities or—even better—the number of students specified by type of education.

Another set of social variables would characterize the *income distribution.* He was concerned about the negative effects of too uneven distributions but hoped that "Models … may soon be used to study in a more precise way … the possibilities to change income distribution and, for example, to find out what degree of inequality is inevitable." Surprisingly enough, he does not refer to the Gini index.

A third set of social variables mentioned would be the sets and costs of *economic and social institutions.* What Tinbergen seemed to have in mind here was not only central and local government agencies but the organization of economic and social life in general, such as it might differ under capitalism or state socialism. As an example, he mentions the costs of tax collection. Had he lived today and seen the mushrooming modern welfare state in all Western countries, he would no doubt have added to his list the cost of operating welfare systems, administrative costs, and perhaps even costs in terms of lost incentives.

What about the overarching preference or utility function of his earlier works? It was still very much in his thoughts. "It is better first to specify the social welfare function as precisely as possible," he said, "and then … use it for finding the socioeconomic optimum." As he concluded his speech, he expressed the hope that one day comparative studies of the socioeconomic optimum would lead to "a discussion of a more scientific character … about the relative merits of various existing socioeconomic orders, especially those of Eastern and Western Europe." He ended by suggesting that "more fundamental research in economics deserves relatively more attention and resources than the more superficial versions of economic research directed at forecasting and analyzing very short-term fluctuations …"

2.3 Goals of economic and social policy

Returning now to the modern problem of organizing economic and social policy to meet the SDGs, we obviously face a situation almost ideal for applying Tinbergen's theoretical framework. In its SDG policy, each country has available a finite list of policy variables—Tinbergen's "instruments," aiming for a number of SDG "targets." Mathematically, Tinbergen asks us to maximize an overriding "social welfare function" like (Eq. 2.2) which in the subsequent will be referred to as the "SDG welfare function." Denote the SDG indicators:

$$r = 1, 2, \ldots, s \tag{2.3}$$

and write the achievements:

$$Y = (Y_1, Y_2, \ldots Y_s) \tag{2.4}$$

Each Y_r element is a single-attribute cardinal number. Next, a set of weights μ_r, $r = 1, 2, \ldots, s$ are defined to be used calculating an index of the total SDG achievements of the country:

$$V = \Sigma \mu_r Y_r \tag{2.5}$$

The index (Eq. 2.5) is a weighted amalgam of the observed SDG indicator achievements. It is the SDG welfare function. The idea of forming such an index is standard in multicriteria analysis. Other commonly used names would be the "scorecard" for the nation examined, or the number of "points" scored.

A well-publicized and ambitious example of the calculation of an index like (Eq. 2.5) is the World Economic Forum *Index of Global Competitiveness* (2011). It aims at measuring the economic, social, and environmental competitiveness of 142 countries. The index is a weighted average of many components, each measuring a different aspect of competitiveness, culminating in a final competitiveness score for each country.

More generally, and in the time-honored language of economic theory, we may see (Eq. 2.5) as a particular (linear) form of social preference or *social utility* obtained in each country, say:

$$V = V(Y) \tag{2.6}$$

While the linear expression (Eq. 2.5) is attractive and easy to work with, there is something to be said for a slightly weaker assumption, only assuming that (Eq. 2.6) is concave. In practical terms, the curvature of the function would measure how a country could compensate for some lack of attainment of one goal by providing more of some other goal. In educational policy for instance, it may be considered to be socially acceptable to compensate for a shortfall of university entrance openings by offering more vocational schools. Differentiating (Eq. 2.6), it would then be possible to determine the exact amount of marginal improvement of one goal that is required to offset a given shortfall of another.

For an illustration, turn to the *map of indifference* or *iso-utility curves* as commonly used in economic textbooks; along each such curve $V = constant$. In Fig. 2.2, two goods are exhibited: Good 1 is measured along the

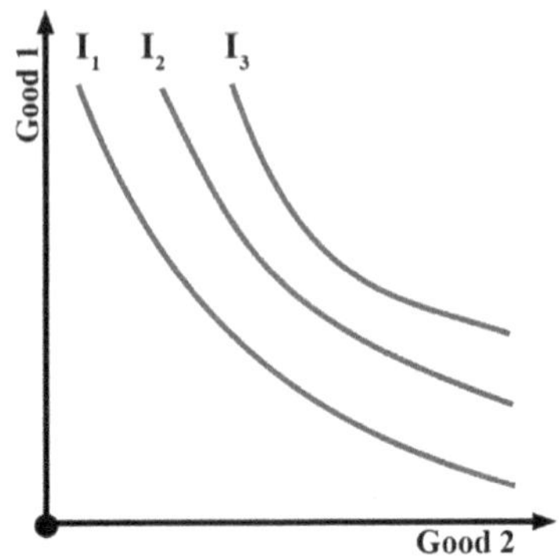

Figure 2.2 Map of indifference or iso-utility curves for the two goals: Goal 1 and Goal 2.

horizontal axis and Good 2 along the vertical axis. In our present application the two "goods" are identified as goals: Goal 1 and Goal 2. The *marginal rate of substitution* between the two goals can be read from the diagram as the slope of an indifference curve. For most goals, the marginal rate of substitution is not constant, so their indifference curves are convex to the origin, describing a negative substitution effect: as the coverage of one goal is lowered, the policymaker needs to be compensated by a greater achievement in terms of the other goal.

An example of a utility function that generates indifference curves that are convex toward the origin is the Cobb–Douglas function. For the calculations to follow, we assume that the SDG welfare function *V(Y)* as spelled out in (Eq. 2.6) is generalized Cobb–Douglas, so that the relationship between social preference and the goals is linear in the logarithms of the variables:

$$\log V(Y) = \sum_{r} \mu_r \log Y_r \tag{2.7}$$

where the μ_r, $r = 1, 2, \ldots, s$ are weights to be determined empirically (a possible constant term is suppressed).

Needless to say, the Cobb–Douglas formulation is still very much within the Tinbergen tradition (see e.g., his Econométrie, (Tinbergen, 1941)).

2.4 The policy parameters

The task of economic and social policy is not only to formulate goals but more importantly to find ways of realizing those goals within the constraints set by existing conditions. Those constraints are not just budget constraints but all sorts of political constraints as well. There is a "cost" involved in every mix of policy action, not just in economic terms but more generally a *political opportunity cost.* The opportunity foregone referred to in this expression is the opportunity of an idealized optimal or *equilibrium* array of policy actions (rather than the possible shortcomings of actual policy). The "opportunity" is the opportunity of optimal planning rather than the reality of actual planning. In other words, the political opportunity cost brings in a *comparison* between idealized actions and actual policy actions. Denoting the list of $i = 1, 2, \ldots, m$ policy inputs by the column vector $X = (X_1, X_2, \ldots, X_m)$, we identify the political opportunity cost as:

$$U(X) \tag{2.8}$$

We have already introduced a map of indifference curves for total economic and social welfare $V(Y)$—see Fig. 2.2. Now let us use the same diagram to illustrate the indifference curves of social policy costs. The two "goods" in the figure are then identified as Policy instrument 1 and Policy instrument 2. This time draw a map of indifference curves of the policy parameters X. Along each such indifference curve, the political opportunity cost $U = constant$. The slope of the indifference curve at a given point indicates the rate of substitution between the policy actions X. These curves will also be taken to be convex toward the origin, indicating that along a given indifference curve, a decreasing use of one policy parameter must be compensated by ever increasing amounts of one or more other policy parameters. For a practical example brought from the field of crime prevention, consider the use of alternative sentences by the courts such as community service and incarceration. In this case, the policy cost to society is not only the cost of administration and cost of operating prisons but also the risk of renewed crimes in the future. Presumably, it should be possible to compensate for some shorter prison sentences by extending the practice of compulsory community services, but there are obvious limits to such practices. Just as we have assumed actual economic and social welfare to be Cobb—Douglas, we shall also assume that the political cost of that welfare is Cobb—Douglas, writing:

$$\log U(X) = \sum_i v_i \log X_i, \tag{2.9}$$

where the v_i, $i = 1, 2, \ldots, m$ are weights to be determined empirically (again, a possible constant term is suppressed).

The cognitive map illustrated in Fig. 2.1 can now be spelled out in more detail as in Fig. 2.3, featuring explicitly the instruments of economic and social policy X_i, $i = 1, 2, \ldots, m$ and the goals Y_r, $r = 1, 2, \ldots, s$.

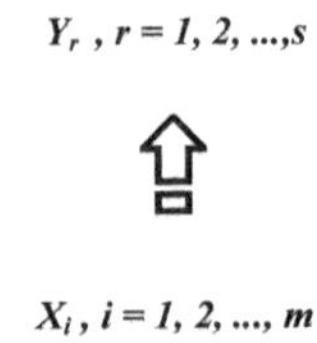

Figure 2.3 Cognitive map illustrating causal relations between policy parameters and goals.

2.5 The effectiveness of economic and social policy

The virtual social cost $U(X)$ should be compared with *actually attained* total societal welfare $V(Y)$. For the purpose of such comparison, form the ratio:

$$V(Y)/U(X) \tag{2.10}$$

which we shall identify as the *effectiveness rating* of the chosen policy. Obviously, the aim of economic and social policy in a country is to make this ratio as large as possible—to maximize the economic and social preference function for any given political opportunity cost.

The idea of forming a ratio like (Eq. 2.10) goes back to standard concepts in engineering where the effectiveness of a machine would be calculated the ratio between actual energy delivered and total energy spent. It would be less than one; only under idealized conditions would it equal one. In neoclassical economic theory based upon axioms of rational behavior, the effectiveness rating would similarly equal one: actual value created is exhausted by costs. But in the real world that we shall be concerned with in our numerical calculations, economic plans sometimes fall short of the possibilities. Opportunities are missed and errors are made. There is then a discrepancy between actual policy and the optimal ideal. The effectiveness ratio (Eq. 2.10) falls below one. The numerical value of the ratio then serves as a metric to indicate the relative success of actual economic and social policy in a country.

Employing the logarithmic specifications of U and V, we shall subsequently choose to define the effectiveness rating of the chosen policy by the ratio:

$$\log V(Y)/\log U(X) \tag{2.11}$$

requiring that:

$$\log V(Y)/\log U(X) \leq 1 \tag{2.12}$$

Again, the effectiveness ratio (Eq. 2.12) may fall below one, signaling a discrepancy between actual policy and the optimal ideal. But if the ratio (Eq. 2.12) equals one, the vector of economic and social policy is optimal, generating goals providing the maximal potential sustainability.

Before turning to the numerical calculations of the effectiveness rating (Eq. 2.12), we need to pause for a moment considering the wider context of the idea of determining a policy effectiveness rating for each nation. What would it mean that some nations reach a desired SDG optimum while others fail to do so? Some nations being optimal but others turning out to be suboptimal (displaying an SDG deficit)? Answer: It means that we are leaving the realm of conventional classical equilibrium economics, entering the uncharted territory of optimization failures.

While Tinbergen in the early 1950s wrote his pathbreaking treatises on economic policy, a consolidation of the inherited economic theory occurred—the codification of neoclassical theory. A generation of young mathematical economists such as Paul Samuelson, Ken Arrow, Gerard Debreu, and Tjalling Koopmans were to formalize the building blocks of an impressive mathematical idealization of a free market economy. The economic agents in the neoclassical economy are abstract consumers and producers, suspended in time, making the rational decisions of "economic man." Markets are in equilibrium.

As we try to understand the arrival of the SDGs on the economic and social scene, obviously some reality checks are in order. Nations do not make rational policy decisions all the time. Real-world nations do not always behave like an "economic man." Planning failures and policy failures do occur. SDG deficits do occur.

Next, consider no longer a single country but a set of countries $j = 1, 2, \ldots n$. Denote the country currently to be rated by the index $j = 0$. The optimization of its rate of policy effectiveness would then require:

$$\text{Max}\left(\sum_r \mu_r \log Y_{r0}\right) / \left(\sum_i \nu_i \log X_{i0}\right) \tag{2.13}$$

Note that the optimization now to be undertaken reaches beyond the scope of the kind of economic and social policy that Tinbergen had in mind. Tinbergen's theory was essentially developed for one single country. He wanted to maximize the "utility function" of the Netherlands. We have indeed formulated a maximand (Eq. 2.13) for a single country $j = 0$, calculated as the ratio between its SDG "utility" and its imputed GDP cost. The constraint (Eq. 2.12) would in the first instance also apply only to country $j = 0$. Now, however, we introduce the additional condition that, still employing the weights determined for nation $j = 0$ and using the *very same* weights for every other nation j, the resulting effectiveness ratio should still remain meaningful:

$$\left(\sum_r \mu_r \log Y_{rj}\right) / \left(\sum_i v_i \log X_{ij}\right) \leq 1, \quad j = 1, 2, \ldots, n \tag{2.14}$$

In words, we want to determine the virtual weights μ and v so that the effectiveness ratio (Eq. 2.12) for the nation currently being evaluated becomes as large as possible, while seeing to it that, using the same weights and constants to calculate the effectiveness ratio of every other nation, all these ratios are meaningful, that is, none of them is permitted to exceed 1. This may sound like a quite obvious requirement but is actually strong enough to solve the entire optimization program, not just for one country but for all countries considered.

Returning to Tinbergen's dictum in his Nobel lecture (see Section 2.4 above), we have indeed "specified the social welfare function as precisely as possible" and we are using it "for finding the socioeconomic optimum." The optimum for each and every country.

Each nation is certainly permitted to have its own SDG priorities (reflected by the weights μ) and its social and economic policy priorities (reflected by the weights v), but they may not deviate from some common consensus to the extent that other nations finds them non-meaningful. The constrained optimization problem (2.13-14) is a case of fractional programming, both the optimand and the constraints featuring fractions. The key to solving it is making the following simple observation. The solution of the unknown weights $(\mu_r,\ r = 1, 2, \ldots, s;\ v_i,\ i = 1, 2, \ldots, m)$ cannot be uniquely determined since the ratio (Eq. 2.13) remains unchanged if we multiply all the weights by a fixed constant. Doubling all the weights, for instance, will leave the ratio unchanged. In what follows, we shall take advantage of this property by normalizing the weights, imposing the condition:

$$\sum_r \mu_r \log Y_{r0} = 1 \tag{2.15}$$

In words, the index of SDGs obtained by weighting (the logarithm of) each goal is normalized to equal 100 per cent. Note that the effectiveness rating then equals the inverted value of the social cost of the bundle of policy instruments. If the effectiveness rating equals 1, it follows that the SDG welfare reached and the social cost of the policy instruments are both equal to 1. The following two propositions then follow:

- If a collection of goals and policy instruments is effective (the rating being equal to 1), then no goal can be increased without lowering the fulfillment of at least one other goal.
- Similarly, the use of no policy instrument can be lowered without increasing at least one other instrument.

These are statements that will ring familiar to any student of neoclassical equilibrium theory, evoking the well-known properties of Pareto efficiency. In order not to mix the concepts, however, we are deliberately using the term "effective" rather than "efficient" to indicate that these are properties of variables that transcend the inputs and outputs of narrow production theory. Collecting all the math, we have now obtained the fractional program (writing it this time with vector notation throughout):

$$\begin{aligned}&\text{Max}\mu\log Y_0/\nu\log X_0\\&\text{subject to}\\&\mu\log Y_j/\nu\log X_j \leq 1, j=1,2,\ldots,n\\&\mu\log Y_0 = 1, \mu,\nu \geq 0.\end{aligned} \tag{2.16}$$

A linear programming equivalent is immediately obtained as:

$$\begin{aligned}&\text{Min }\nu\log X_0\\&\text{subject to} - \mu\log Y_j + \nu\log X_j \geq 0, j=1,2,\ldots,n\\&\mu\log Y_0 = 1, \mu,\nu \geq 0.\end{aligned} \tag{2.17}$$

Whereas we in Chapter 1 touched only lightly on some of the technicalities of our mathematical machinery, they now have to be faced upfront. This necessitates a full discussion of the operations research technique of *data envelopment analysis* (DEA).

2.6 Data envelopment analysis

For most readers of these pages, the disequilibrium technique of DEA will be unknown. DEA has been used extensively during the last 30 years in the estimation of production functions for manufacturing corporations, hospitals, commercial banks, and electric utilities. Some of these units can be represented as profit maximizers, but others cannot. DEA works just as well for a study of farming in a developing country in which most of the farm produce is consumed on the farm, as it does for US farms in the wheat belt. It works just as well for a study of hospitals in Great Britain (which has a socialized health care system) as it does for profit-maximizing hospitals in the United States. For some such units, there may be decreasing returns to scale, as required by standard equilibrium theory. For others, there may be rapidly increasing returns to scale, even hyper-growth (in the semiconductor or computer industries).

In all of these applications, the task of DEA is to identify a production *frontier* and to determine the location of each observation (hospital, bank, and electric utility) relative to the frontier. On-frontier observations are ascribed an efficiency rating of 100%, off-frontier observations efficiency ratings less than 100%. For each off-frontier observation, the "distance" (shortcomings) to the frontier is calculated and thus the scope for possible improvement is determined.

Using DEA to rate the economic and social policy of a group of entire countries, the performance or effectiveness of each country will be assessed on the basis of its ability to realize the list of policy goals laid down. Again, frontier observations will be ascribed an effectiveness rating of 100%. Geometrically, the frontier is constructed as the envelope of the 100% effective countries. In modern hi-tech lingo, the word envelope is used in expressions such as "pushing the envelope," meaning going beyond what is technically possible today. More strictly, envelope is a geometrical term, referring to the piece-wise linear locus of tightest closure around a given collection of points. Off-frontier countries are ascribed an effectiveness rating of less than 100%. These countries fall behind (or below) the envelope. Projecting an off-frontier country onto the envelope, one finds its ideal performance. The difference measures the potential for possible improvement of economic and social policy.

The surprising connection between Tinbergen's idea of maximizing a social welfare function and DEA was discovered fairly recently, in a 2010 paper by Cooper, Thore, and Tarverdyan printed in the volume *Global Operations Management*, ed. L.R. Hockley.

2.7 Constructing the frontier

A simplified illustration of an economic and social policy frontier is provided in Fig. 2.4. The achievements of six countries A, B, C, D, E, and F are compared, plotting for each country the use of a single policy instrument X (an "input") and the resulting performance in terms of a single sustainability goal Y (the "output"). The observation for each country is marked as a separate point.

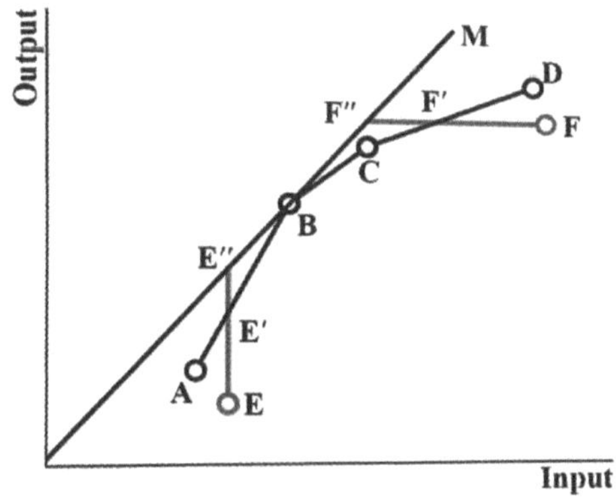

Figure 2.4 Illustration of instrument—goal effectiveness frontier.

Two different envelopes can be defined. One is called the "the upper convex hull" to the observed points. In Fig. 2.4, it is represented by three linear facets: AB, BC, and CD. The frontier ABCD is said to be "spanned" by the four countries: A, B, C, and D. They are located on the instrument—goal effectiveness frontier. They are effective. The two remaining countries E and F are located behind the frontier. They are ineffective.

The other envelope is called "the conical hull" to the given points. It consists of ray O-B-M drawn from the origin O, touching only point B. Along this envelope, there are throughout constant returns to scale Y/X. By contrast, the convex cone A-B-C-D displays variable returns to scale, increasing returns to scale along AB and decreasing returns to scale along BC and CD.

In this book we shall mainly use the constant returns to scale alternative. It forces the supporting hyperplanes to pass through the origin, thus reducing the number of observations rated as effective. On the other hand, if the countries are fairly similar, one may prefer to use the variable rate of returns to scale envelope, thus establishing a somewhat tighter fit. More countries will then be rated as effective (The calculations reported in Chapter 1 assumed variable returns to scale.).

When the inputs and goals are all multidimensional, the envelope takes the form of a linear polyhedron. In the case of the upper convex hull, it consists of planar facets and corner points. The corner points span the envelope. The conical hull also consists of planar facets and corner points, this time located on supporting hyperplanes running through the origin.

Historically, the envelope O-B-M with constant returns was applied in the pioneering paper of Charnes et al. (1978). For the first few years, the

CCR model (named after the initials of the authors) was the only enveloping procedure around. But with the publication of the Banker et al. (1984) paper—later referred to as the BCC model—the alternative with flexible returns to scale (A-B-C-D) also became investigated.

The envelopment provides a ranking of all ineffective countries (There is no ranking of the effective units because they all receive the rating of 100%. They are "the best."). An effectiveness score less than 100% signals nonoptimal and nonequilibrium behavior. The performance of such a country is non-Paretoan in the sense that, given the policy instruments X, it should be possible to increase the attainment of one or more goals Y. Fitting the upper convex hull in Fig. 2.4, the effectiveness ratings of countries A, B, C, and D are all 100%; all countries below A, B, C, and D will be assigned a rating of less than 100%. Fitting a conical hull instead, the effectiveness rating of point B is 100% and the rating of each point below OB is less than 100%.

The observations for a given suboptimal country can be "projected" onto the frontier, keeping its policy inputs fixed but increasing its goal achievements equiproportionally. In this manner, country E can be projected onto the variable-returns-to scale envelope at point E'. The projection onto the constant-returns-to scale envelope is point E''. Such projections are referred to as "best practice."

Best practice can also be obtained as a weighted average of the practices of two or more effective countries (located on the frontier), called the "peers." It is a linear "amalgam" of the performance of those peers.

Thus, DEA is not only diagnostic in that it identifies those practices that perform optimally, but it is also normative or prescriptive in that it delivers information how the actual performance of a suboptimal practice might be improved. Assessing the performance of a collection of countries, we shall be able to:

- identify and separate the frontier and subfrontier countries,
- identify the peers of the subfrontier countries,
- pinpoint the shortcomings of lagging countries and provide advice for corrective policy action.

2.8 The mathematics of data envelopment analysis

We now propose to fit the frontier or envelope to the logarithmic data:

$$\left(\log X_j, \log Y_j\right),\ j = 1, 2, \ldots, n \tag{2.18}$$

where, as before, X_j is the vector of policy instruments employed by country j and Y_j is the vector of goals. The task at hand is to rank country $j = 0$ relative to the frontier to be defined. The policy instruments and the goals achieved by this country are:

$$(\log X_0, \log Y_0) \tag{2.19}$$

Let λ_j, $j = 1, 2, \ldots, n$ be nonnegative weights attached to each country. The weights are unknown, to be determined. Form the "composite country" with the frontier policy costs $\Sigma_j(\log X_j)\lambda_j$ and frontier goal achievements $\Sigma_j(\log Y_j)\ \lambda_j$.

Ranking country $j = 0$, if it so happens that the weight of this country is one and the weights of all other countries are zero, then the country will be said to lie on the frontier. A few countries will always be located on the frontier, and the frontier is defined by or "spanned" by these frontier countries.

In order to find out whether a given country $j = 0$ is located on the frontier, or behind it, the following procedure is introduced. One attempts to expand its goal achievements from the actual values $\log Y_0$ to an expanded level $\Phi \log Y_0$ where Φ is an expansion factor to be determined. The expansion can continue as long as the attained level stays below the frontier goal, but no further:

$$\Phi \log Y_0 - \Sigma_j(\log Y_j)\lambda_j \leq 0 \tag{2.20}$$

At the same time, frontier policy costs cannot exceed actual costs:

$$\Sigma_j(\log X_j)\lambda_j \leq \log X_0 \tag{2.21}$$

The expansion now described can be described as "projecting" the current country onto the frontier. Bringing all the math together in a single linear program:

$$\begin{aligned}
&\text{Max } \Phi \\
&\text{subject to} \\
&\Phi \log Y_0 - \Sigma_j(\log Y_j)\lambda_j \leq 0 \\
&\Sigma_j(\log X_j)\lambda_j \leq \log X_0 \\
&\Phi \text{ unconstrained}, \ \lambda_j \geq 0, \ j = 1, 2, \ldots, n
\end{aligned} \tag{2.22}$$

The corresponding dual program reads:

$$\begin{aligned}
&\text{Min } \nu \log X_0 \\
&\text{subject to} \\
&-\mu \log Y_j + \nu \log X_j \geq 0, j = 1, 2, \ldots, n \\
&\mu \log Y_0 = 1, \mu, \nu \geq 0
\end{aligned} \tag{2.23}$$

where μ is the dual to (Eq. 2.20) and ν is the dual to (Eq. 2.21). Note that program (Eq. 2.23) is identical to program (Eq. 2.17), which proves that the frontier determined by maximizing a Tinbergen-type SDG welfare function is indeed the envelope as determined by DEA. *Quod erat demonstrandum.*

If the constraint $\Sigma_j \lambda_j = 1$ is adjoined to the program, one obtains a variable-returns-to scale model (Banker et al., 1984); if it is omitted one gets the constant-returns-to-scale model (Charnes et al., 1978).

Finally, the concept of *slack* is important. Even if the expansion factor Φ comes out as equal to one, there may still exist one or several goals that fall short of the best practice goal, i.e., there is positive slack in one or several of the constraints $\log Y_0 - \Sigma_j (\log Y_j)\lambda_j \leq 0$. Similarly, there may also exist one or several policy instruments for which best practice falls short of actual use, i.e., there is positive slack in one or several of the constraints $\Sigma_j (\log X_j) \lambda_j \leq \log X_0$. For a country to be rated as effective, however, we shall not permit any such positive slack. An effective country, then, needs to fulfill two conditions:

- The effectiveness rating equals one,
- The country exhibits no positive deviations from best practice (neither a goal slack nor a policy instrument slack).

Only then will we have made certain that the country is actually located on the envelope itself.

Denoting the goal slacks $\Sigma_j (\log Y_j)\lambda_j - \log Y_0$ by the letter s and the policy slacks $\log X_0 - \Sigma_j(\log X_j)\lambda_j$ by the letter t, program (Eq. 2.22) then will be modified to read:

$$
\begin{aligned}
&\text{Max } \Phi + \varepsilon(s + t) \\
&\text{subject to} \\
&\Phi \log Y_0 - \Sigma_j\left(\log Y_j\right)\lambda_j + s = 0 \\
&\Sigma_j\left(\log X_j\right)\lambda_j + t = \log X_0 \\
&\lambda, s, t \geq 0 \\
&\Phi \text{ unconstrained}
\end{aligned}
\tag{2.24}
$$

where ε is a small so-called non-Archimedean number. It effectively allows the maximization over Φ to preempt the optimization over the slacks. Thus, the optimization can be computed in a two-stage process with maximal expansion of the goals Φ being achieved first; then at the second stage, movement onto the efficient frontier is achieved via the slack variables.

CHAPTER THREE

Before and after the pandemic: a dashboard of sustainable development goal metrics for assessing individual well-being

The original purpose of the present book was a straight-forward one: to document how nations more or less successfully use macroeconomic policies to further their achievement of the UN Sustainable development goals. A clear research task calling for the multidimensional programming techniques outlined in Chapters 1 and 2. And then, in January 2020, the COVID-19 virus left the Chinese mainland and started its death march throughout the world.

Most Western nations chose to tighten their borders, introducing various measures of social distancing, lockdown, closing schools and restaurants. It seemed to work for a while during late spring 2020 and the summer, but with disastrous consequences on employment and business activity. In the fall, the sentiment was for easing the restrictions. At the same time, the number of cases started to rise again, threatening a "second wave" of COVID-19. In different ways, all countries faced the same problem: how to control the policy levers of economic and social policy while at the same time limiting the risk of infections and death. Should schools be opened again or stay closed? Should care homes stay in lockdown?

The theme that we have chosen for the present chapter is Individual Well-being. By that term we mean not only health and education (the five first sustainable development goals [SDGs]) but also living conditions and individual well-being in a wider sense including a safe and clean home, sanitation (SDG6), waste collection (SDG12), and access to social services. We make this choice in order to establish a base point for the subsequent analysis of the COVID-19, measuring the degradation of individual well-being caused by the pandemic.

The UN Statistical Commission had originally listed 169 indicators of the SDG achievements of nations. The sheer number of dimensions involved means that no final and conclusive study of the SDG ramifications on national economic and social policies can ever be made. Instead, we have

Measuring Sustainable Development Goals Performance
ISBN: 978-0-323-90268-7
https://doi.org/10.1016/B978-0-323-90268-7.00007-4

to pick and choose indicators that seem relevant for (shedding light on) the problem at hand. The term "dashboard" well captures the idea of a broad set of metrics available to guide the conduct of policy. The SDGs taken all together constitute a huge dashboard, but only some subset of the available metrics is needed for any particular policy task.

Focusing our work on Individual Well-being, we unfortunately shall have to leave aside the many sustainability goals relating to the climate, pollution, the loss of biodiversity burning issues that all deserve careful study, but which we shall have to leave to other researchers.

3.1 Dimensions of individual well-being

Selecting the final list of indicators to be used for the computer runs, it emerged quickly that the apparent wealth of available indicators is mainly an illusion, the main reason being the problem of missing data. For the optimization programs outlined in Chapter 2, one needs observations of each selected goal indicator and each selected policy instrument indicator, for every single country. Alas, the World Bank and the Organization for Economic Cooperation and Development (OECD) data banks too often report missing data for one or several countries.

Eventually, we ended up with the following seven indicators of individual well-being:

- *OBESITY*—an indicator of SDG II, defined as individuals having a body mass index of 30 or more; in % of the adult population (OECD code name "sdg2_obesity"), latest data for 2016.
- *SANITA*—an indicator of SDG VI, % of population using at least basic sanitation services ("sdg6_sanita"), latest data for 2017.
- *UNEMPL*—an indicator of SDG VIII, unemployment rate, % of total labor force ("sdg8_unemp"), latest data for 2019.
- *MOBILE*—an indicator of SDG IX, the number of mobile broadband subscriptions per 100 inhabitants ("sdg9_mobuse"), latest data for 2018. The mobile phone and the availability of broadband subscriptions at a reasonable price is rapidly changing the lives of ordinary people.
- *ELDERLY*—an indicator of SDG X, the elderly poverty rate calculated as % of population aged 66 or over ("sdg10_elder"), latest data for 2016.
- *WASTE*—an indicator of SDG XII, amount of municipal solid waste in kg/day/capita ("sdg12_msw"), latest data n.a.
- *SAFE*—an indicator of SDG XVI, % of population feeling safe walking alone at night ("sdg16_safe"), latest data for 2019.

As noted earlier, some of the UN goals are desirables, others are unwanted. But for programming purposes, we needed to convert *OBESITY, UNEMPL, ELDERLY,* and *WASTE* to something wanted rather than something unwanted. This is easy when the indicators are simple percentage points. For *WASTE* we picked the number 10 kg/day/capita as an arbitrary upper limit on the amount of discharge of municipal waste. Altogether then, we formed the desirables 100 − *OBESITY, SANITA, 100 − UNEMPL, MOBILE, 100 − ELDERLY, 10 − WASTE,* and *SAFE.*

The computer runs reported in the present chapter all refer to the member states of the OECD, Paris. The data were obtained from the websites worldbank.org and data.oecd.org. The most recent available data up to and including 2019 were used throughout. That year the OECD had 36 member states, as listed below:

Australia, Austria, Belgium, Canada, Chile, Czechia, Denmark, Estonia, Finland, France, Germany, Greece, Hungary, Iceland, Ireland, Italy, Israel, Japan, Korea, Latvia, Lithuania, Luxembourg, Mexico, Netherlands, New Zealand, Norway, Poland, Portugal, Slovenia, Slovakia, Spain, Sweden, Switzerland, Turkey, UK, US. (3.1)

The European nations in OECD include 25 EU nations (Malta and Romania are not members of the OECD) plus Switzerland.

3.2 Successes and failures

Each year, the secretary-general of the Economic and Social Council reports on progress toward the SDGs. In his 2020 report[1], he noted that progress had continued in some areas: poverty continued to decline, albeit at a slower pace; maternal and child mortality rates were reduced; more people gained access to electricity. In other areas, however, progress had either stalled or been reversed: the number of people suffering from hunger was on the rise, and climate change was occurring much faster than anticipated.

In particular, the report highlights the impacts and implications of the COVID-19 pandemic on several SDGs. Here are some of its conclusions:

Combatting poverty (SDG 1): The pandemic is reversing earlier progress in poverty reduction. An estimated 40−60 million of people will be pushed

[1] Progress toward the Sustainable Development Goals, Economic and Social Council, UN, New York 2020.

back into extreme poverty, causing the first increase in global poverty in more than 20 years. About four billion people do not benefit from any form of social protection, which is critical to help the poorest and the most vulnerable in this crisis.

Ending hunger (SDG 2): The total number of people suffering from hunger is on the rise and there are millions of malnourished children. The economic slowdowns and disruptions in the food value chains caused by the pandemic will exacerbate hunger and food insecurity.

Ensuring healthy lives (SDG 3): The COVID-19 pandemic is devastating health systems globally.

Education and life-long learning (SDG4): Southern Asia is home to almost half of the global illiterate population and sub-Saharan is home to one quarter.

Gender equality (SDG 5): The current pandemic is hitting women and girls hard. Globally, women make up three quarters of medical doctors and nursing personnel.

Water and sanitation for all (SDG 6): In 2017, three billion people lacked soap and water at home. Close to half of the schools in the world lack hand-washing facilities.

Notice that in all these cases, the SDGs are seen as cause variables of the virus rather than its effects: poverty, hunger, bad hygiene, or poor access to clean running water facilitates the spread of the virus. The causality only weakly runs in the opposite direction, such as poverty becoming more common in infected regions, or the availability of nurses and carers becoming more scarce as many of them catch the virus themselves. On the other hand, there are many examples of the virus prompting the rapid construction of new hospital facilities (such as the assembly of the Nightingale hospitals in England and Scotland, and Northerns Ireland) or the startup of mass production of respirators and personal protection equipment, all actually improving the *SDG 3* achievements of a nation.

The Economic and Social Council report also discusses the current prospects of economic development and the bleak outlook for meeting many of the SDGs.

Reducing inequality (SDG 10): In many countries with comparable data during the period 2012–2017, the bottom of the population saw their incomes grow. Still the top 10% of the population received at least 20% of the income.

Safe cities and human settlements (SDG 11): Rapid urbanization is resulting in a growing number of slum dwellers, inadequate and overburdened infrastructure and services, and worsening air pollution.

Crowded conditions increase the incidence of the virus.

3.3 The political impact of the sustainable development goals

While the record of achievements of the SDGs is a mixed one, the concept of sustainability itself turned out to be a true game-changer. Already the Brundtland report[2] had pointed at the need to see ecological stress—degradation of soils, water regimes, atmosphere, and forests—as a trend that threatens to radically alter life of the species on the planet, including the human species. The various global "crises" that seize public concern should not be seen as separate crises. The environmental crisis, the development crisis, and the energy crisis are all interlocking, they are all one: a crisis of sustainability.

The wave of more than one million refugees and immigrants entering Europe from Syria, Iraq, and Afghanistan in 2015 was another signal event during the time period that we are reviewing here. The views of the public were divided about the diversity represented by these immigrants. Experience has shown that many of the new arrivals were eventually well assimilated in their new countries. In other cases, the influx led to the appearance of urban ghettoes riddled by crime (*SDG 11*) as in Manchester or Malmö, Sweden.

A joint report by the OECD and the EU titled *Settling In 2018: Indicators of Immigrant Integration*, Paris, Brussels 2019, argues that failed integration is real. Key findings:

> *In most domains, immigrants tend to have worse economic and social outcomes than the native-born;*
>
> *Over the last decade, differences in unemployment rates of immigrants and native born have widened;*
>
> *Across the OECD half of all children in immigrant households live below the relative poverty line, compared to over a quarter in native born households. The countries with the highest shares of immigrant offspring living in relative poverty are Spain, Greece, and the United States;*
>
> *In the EU, around 15% of all non-EU born aged 15 to 64 went no further than primary school education.*

[2] *Our Common Future*, Report of the World Commission on Environment and Development, Oxford University, Cambridge 1987.

The new millennium has brought growing social and economic problems that may lie at the bottom of the wellness deficits that concern us here. Some of them are listed below.

- Overpopulation: Medical progress in combatting disease has led to rapid increase of the population in large parts of the world. Many of the fastest growing populations are in the poorest countries, where population growth brings additional pressure on the means of subsistence. According to the UN reports, the population of sub-Saharan Africa is projected to double by 2050. Around 2027, India is projected to overtake China as the world's most populous country (see *World Population Prospects, 2019: Highlights,* UN Department of Social and Economic Affairs).
- Falling standards of nutrition: The UN Decade of Action on Nutrition (2016—2025) is a commitment by United Nations member states to undertake 10 years of sustained and coherent implementation of programs and increased investments to eliminate malnutrition in all its forms. The Decade of Action is organized by the WHO and the FAO (recipient of the Nobel peace prize for 2020). The intent is to eradicate nutrient deficiencies, overweight or obesity, and to reduce the burden of diet-related diseases. The program discourages industrial foods bought in the supermarket rather than cooked at home. It discourages working women buying readymade meals for the family, no longer serving meals around a dining table. See the website www.who.int/nutrition/decade-of-action.
- Illiteracy: In a 2019 publication entitled *The Education Crisis: Being in School is not the same as Learning,* the World Bank states that, worldwide, hundreds of millions of children reach young adulthood without the most basic skills like calculating the correct change from a transaction, reading a doctor's instruction, or understanding a bus schedule— let alone building a fulfilling career or educating their children. And at the annual meeting of the World Economic Forum in Davos in January 2020, the World Bank's director of Education Global Practice asked why 53% of 10-year-olds in low- and middle-income countries cannot read and understand a simple text.

The arrival of the new millennium came to signal deep going changes in the cultural makeup of the Western civilization. The signing of the *UN Agenda 2030* in New York in September 2015 was a historical act that could not have occurred 20 years earlier. Back then, there certainly was awareness of social and environmental problems but, by and large, individual nations were left to cope with them on their own. The idea that a supranational authority would intervene in national sovereignty, asking for detailed annual progress reports was mote.

More fundamentally, our understanding of a Western liberal society has evolved to reflect new political agenda. Nowhere is this more striking than in the field of education, with the curriculum in most subjects being rewritten many times in Western nations, emphasizing the right of every individual to take charge of their own lives and to make their own decisions. As a result, we live in a more vociferous world, everybody claiming their human rights.

The idea of multiculturalism expresses the equal value of every human being, rejecting tribal or national patriotic allegiances, hoping to replace them with an international fellowship of nations, no doubt striving for the SDGs. But not every nation is willing to sacrifice the old loadstars. As we shall see from the data work, many East European nations stick to the old ways, remembering the yoke of communism and still inspired by the church.

3.4 The list of policy instruments

Returning for a moment to Tinbergen's idea of maximizing a social welfare function subject to policy constraints, we find it natural to identify those constraints with the availability of standard macroeconomic causative factors such as GDP, and investments in research and development (R&D). Altogether, we used the following three indicators of economic and social policy:

- *GDP* gross domestic product per capita, in 2017 US dollar equivalents (purchasing power parity),
- *RANDD* investments in R&D, in *pro mille* of GDP (several OECD countries report R&D less than 1% of GDP which would make the logarithm negative; hence we opted for pro mille instead), listed by the OECD as an indicator of *SDG IX* with the code name "sdg9_rdex," latest data for 2017,
- *HEALTHED* government spending on health and education, rather curiously listed by the OECD as an indicator of *SDG XVII* with the code name "sdg17_govex," latest data for 2016.

In a fashion, these three factors can be seen as key levers of economic policy that each nation have available to improve the well-being of its citizens. The GDP is a shorthand for all economic activity in a nation; we have already in Chapter 1 explored its impact upon the SDGs. Investments in R&D and government spending on health and education both constitute links to the future, hopefully a future with improved individual well-being.

Investments in R&D are of course a standard economic variable measuring the increase of "human capital." Government spending on health and education is included in our list with some misgivings: government involvement in hospital care and schools varies considerably from one other country to another. In the United States, for instance, education is mainly the responsibility of each individual school district, and paid for by local property taxes. Other countries have other arrangements, raising funds privately or regionally.

Our presumption, then, is that these three factors will be positively associated with the seven indicators of personal well-being. If it turns out that such an association is only weak or even absent, there is cause for worry.

3.5 A first look at the computing results: Pareto optimal OECD nations

For the mathematical work, we return to the optimization program (Eq. 2.26) assuming constant-returns-to-scale (the CCR model). Pareto optimality requires that the expansion factor Φ (being the inverse of the rate of policy effectiveness) equals one, and that all slacks vanish $s = t = 0$.

Solution showed that the following 16 OECD countries were Pareto optimal:

Chile, Estonia, Finland, Greece, Hungary, Japan, Korea, Latvia, Luxembourg, Mexico, Poland, Portugal, Slovenia, Slovakia, Spain, Turkey. (3.2)

These nations all reached a policy effectiveness rating of one, exhibiting neither input nor goal slack. They all achieved maximal SDG goal results. Inspecting the list, we note that it includes several Eastern European nations. But some large Western European nations are missing. France, Germany, the UK, and Italy did not make it.

For the nations (Eq. 3.2), their "best practice" or frontier practice coincides with actual practice, see Table 3.1 below.

Remember that the optimization solves for an entire team of countries defining the envelope circumscribing the results of all the nations. It does not solve for individual world winners (Japan would be the gold medalist in terms of 100 − OBESITY, SANITA, 100 − UNEMPL, MOBILE, Slovakia in terms of 100 − ELDERLY, Poland in 100 − WASTE and Slovenia in SAFE). Instead, it solves for the joint group of "model" countries that collectively define the performance frontier. In that sense, their achievements are all equally good. They are all available as a team to serve as models

Table 3.1 Sixteen Pareto optimal OECD countries. All goals measured in original scales. For each of these countries, the observed goal achievement coincides with the calculated optimal one.

	OBESITY	SANITA	UNEMPL	MOBILE	ELDERLY	WASTE	SAFE
Chile	28.00	99.89	7.01	88.25	16.30	1.08	47.39
Estonia	21.20	99.61	5.59	133.35	35.70	1.47	74.89
Finland	22.20	99.45	8.25	153.76	5.00	2.13	87.57
Greece	24.90	98.96	21.08	63.40	7.80	2.00	57.49
Hungary	26.40	97.99	3.65	63.19	8.60	1.92	63.55
Japan	4.30	100.00	2.44	180.95	19.60	1.71	75.73
Korea	4.70	99.89	3.73	112.77	45.70	1.24	72.89
Latvia	23.60	92.86	8.57	116.12	32.70	1.03	61.66
Luxembourg	22.60	97.61	5.36	88.10	7.70	2.31	83.53
Mexico	28.90	89.22	3.37	62.79	25.60	8.76	39.67
Poland	23.10	98.13	4.36	154.12	9.30	0.88	66.12
Portugal	20.80	99.44	7.35	68.88	9.50	2.21	75.79
Slovenia	20.20	99.11	6.22	70.03	12.30	1.21	90.43
Slovakia	20.50	98.93	7.27	82.64	4.30	1.37	66.70
Spain	23.80	99.90	14.55	94.11	9.40	2.13	84.05
Turkey	32.10	96.37	11.14	70.52	17.00	1.77	56.84

or "peers" to the suboptimal nations falling behind the envelope. Some are more popular peers than the others, Poland serving as peer to 17 suboptimal nations, Portugal to 12, Mexico to 5, and Estonia to 4 suboptimals. The team of peers consists of 13 active members. and three sleeping members who currently do not serve as role models to anybody else: Korea, Luxembourg, and Slovakia.

Which are the characteristics that are required for a country to be recognized as a frontier point and enter the select club of Pareto optimal nations? What exactly does it take to excel in terms of those seven indicators of personal wellness that we have identified?

Here is a quick snapshot of individual wellness in Portugal, the second most cited example of the peers: The industrial revolution came relatively late to this country and it never brought the kind of industrial slum that mars many other nations. Most Portuguese can trace their roots to a small farming or fishing village, with the ruin of the ancestral home still standing. Family bonds are strong. The Portuguese are proud of and cherish their glorious cultural heritage. Schools and education are of excellent quality. Migrants from Brazil or from Portuguese-speaking areas in Africa easily meld with society. The recent wave of migrants over the Mediterranean has barely touched Portugal. Crime is low.

Many of the peers (Eq. 3.2) are traditional agricultural societies with moderate incomes (less than US $ 30,000 per capita—exceptions: Finland, Israel, Japan, Korea, and Luxembourg) and comparatively low government spending on health and education (less than 10% of GDP—exceptions: Japan and Finland). But money spent apparently tells only part of the story. The quality of efforts matters more than the expenditure account. Somehow, the frontier countries were able even with limited means to achieve superior results in terms of health, hygiene, and safety.

Dollars by themselves do not produce well-being.

3.6 Disequilibrium: the deficit countries

The following 20 OECD countries turned out to be non-Pareto:

Australia, Austria, Belgium, Canada, Czechia, Denmark, France, Germany, Iceland, Ireland, Italy, Israel, Lithuania, Netherlands, New Zealand, Norway, Sweden, Switzerland, UK, US. (3.3)

This list includes some of the wealthiest and most modern nations in the world. It fits the picture that emerged already from our discussion in Chapter 1: riches in terms of GDP do not necessarily imply superior personal well-being. Apparently, nor does plentiful spending on R&D or health and education. The conclusion seems to be that the dimensions of well-being that we have identified here (healthy nutrition, sanitation, job security, social contacts, safety) do not respond well to macroeconomic policy tools.

How much better would the frontier performance be? Fortunately, we can give quite precise answers to that question. To begin with, let us look at the solution policy effectiveness ratio of each country (obtained as $1/\Phi$ in the math of the programming problem). The ratio is less than 1 for all non-Pareto countries, but some of them are doing better than the others. A few of them actually reached a policy effectiveness rating $1/\Phi$ quite close to 1. Here are the details, the countries this time being ordered by falling rates of policy effectiveness $1/\Phi$:

Czechia 0.995, Lithuania 0.990, Israel 0.989, Ireland 0.987, Denmark 0.983, France 0.980, Iceland 0.975, Italy 0.975, New Zealand 0.975, UK 0.973, Netherlands 0.969, Austria 0.967, Canada 0.965, Sweden 0.962, Norway 0.960, Belgium 0.958, Germany 0.958, Switzerland 0.956, Australia 0.955, US 0.943. (3.4)

The inverse of the effectiveness ratio is the factor Φ showing the rate of equiproportional expansion of all goal achievements that would be necessary to raise the performance of a nonfrontier country to parity with the frontier ones. For the nations at the top of the list, this expansion would be fairly modest. But for the trailing nations, led by the US, a quite considerable increase of the recorded goal achievements would be necessary:

Czechia 1.005, Lithuania 1.010, Israel 1.011, Ireland 1.014, Denmark 1.018, France 1.021, Iceland 1.025, Italy 1.025, New Zealand 1.025, UK 1.027, Netherlands 1.032, Austria 1.035, Canada 1.036, Sweden 1.039, Norway 1.042, Belgium 1.044, Germany 1.043, Switzerland 1.046, Australia 1.047, US 1.060. (3.5)

In order to double-check the results now obtained, we decided also to run a variable-returns-to-scale optimization model (the BCC model) on exactly the same data. This produced the following list of Pareto optimal countries:

Australia, Austria, Belgium, Chile, Czechia, Denmark, Estonia, Finland, France, Greece Hungary, Iceland, Ireland, Israel, Japan, Korea, Latvia, Lithuania, Luxembourg, Mexico, Netherlands, New Zealand, Norway, Poland, Portugal, Slovenia, Slovakia, Spain, Switzerland, Turkey. (3.6)

Only six countries turned out to be non-Pareto:

Canada, Germany, Italy, Sweden, UK, US (3.7)

They all scored a suboptimal policy effectiveness ratio of 0.999. As can be seen, the very same six countries came out as suboptimal also with the constant-returns-to-scale model. As mentioned earlier, the variable-returns-to-scale envelope tends to produce a better fit to the given observations.

The computer solution will deliver not only the optimal policy effectiveness ratio for each country but also detailed information about the performance deficits of each suboptimal country. Each goal deficit is calculated as the difference between the recorded data and "best practice" obtained as the projection of the actual observations onto the frontier.

We provide in Table 3.2 below a listing of "best practice" for each of the six countries (Eq. 3.7). The same table using the constant-returns-to-scale envelope did not produce any meaningful results (for instance, best practice for the SANITA variable then came out as more than 100% for all countries, and best practice for UNEMPL came out as negative for all countries, illustrating the poor fit of the envelope).

Each one of these six countries present an individual wellness deficit in terms of every one of our seven indicators. Every recorded entry in the table falls short of the frontier performance. For Canada and Germany, the deficits

Table 3.2 Six subeffective OECD nations fitting a Variable-returns-to-scale envelope. In each cell: upper number lists actually observed data, in original scales; lower number is the calculated best practice.

	OBESITY	SANITA	UNEMPL	MOBILE	ELDERLY	WASTE	SAFE
Canada	29.4	98.96	5.87	63.4	10.5	2.33	57.49
	23.5	99.0	5.4	87.4	10.1	2.1	82.5
Germany	22.3	99.22	3.72	79.78	9.6	2.11	74.46
	19.9	99.6	3.3	98.9	9.3	1.71	76.9
Italy	19.9	99.28	10.57	87.91	10.3	2.23	63.71
	19.7	99.6	10.3	101.0	10.0	1.81	75.4
Sweden	20.6	99.30	6.32	122.52	11	1.61	77.1
	17.9	99.6	4.5	122.9	10.8	1.6	77.3
UK	27.8	99.11	4.00	88.06	14.2	1.79	79.33
	22.7	99.13	3.98	115.4	14.18	1.789	79.34
US	36.2	99.97	4.84	132.91	22.9	2.58	72.38
	8.9	100	2,85	165.5	18.3	2.05	74.7

are considerable in terms of human contacts (MOBILE) and personal safety. Italy faces a huge deficit in personal safety. Sweden is doing quite well; it could do better in terms of the prevalence of obesity in its population, and of its unemployment. The UK is not faring well in terms of personal mobile conversations. And the results for the US are overall quite poor.

The frontier calculations also provide information about the "peers" or performance "models" of each suboptimal country. Remember that the peers are those Pareto optimal countries that together form the weighted average that is "best practice." So, which are those peers and how often do they serve in this function? Some peers are more popular than others. For the case of the constant-returns-to-scale envelope, the four most common peers are listed below, together with the number of times each peer serves as a reference to a suboptimal country:

Poland 17 times, Portugal 13 times, Slovenia 13 times, Estonia and Mexico four times. (3.8)

These are the countries that seem to teach the economic heavyweights of Western Europe, Canada, and the US a lesson.

3.7 Asking questions

Our findings are both surprising and worrying. Why do not modern Western nations deliver better on personal wellness? Almost half of all OECD nations display a personal wellness deficit; they include some of the richest and technologically most advanced Western nations.

The shortcomings occur in terms of some key indicators of:
personal health and nutrition (100 − OBESITY),
hygiene (SANITA),
family financial security (100 − UNEMPL),
social contacts (MOBILE),
economic protection for retired persons (100 − ELDERLY),
absence of polluting waste around the home (10 − WASTE),
a safe neighborhood (SAFE).

Is it true that pockets of poverty have opened up in the midst of plenty, not seen or not wanted to be seen by policymakers? And why are culturally more conservative nations doing so much better?

Let us repeat what it means that a nation is found to be non-Pareto. Geometrically, it means that the nation falls below a supporting hyperplane to two or more Pareto observations (its peers). It does not mean that every single goal achievement of the nation falls behind those of its peers.

For instance, comparing the results of the US with those of its major peer Estonia, the goal results of the US in terms of 100 − OBESITY, 100 − ELDERLY, 10 − WASTE and SAFE fall below those of Estonia, but the US got better results for SANITA, 100 − UNEMPL, and MOBILE. So, what we are dealing with here are averages. All indices are averages. The performance index of the US falls below the performance index of Estonia.

Already our investigation of the relationship between GNP and the SDGs in Chapter 1 suggested that many GDP-rich Western nations are falling short of the SDGs. Now we are using a much broader set of macroeconomic variables including R&D and government expenditures on both health and education, presumably all beneficial to social and economic progress, and yet advanced nations like France, Germany, the UK, and the US are falling behind a Pareto effective frontier established by Estonia, Poland, Portugal, Slovenia, and others.

Remember that we are dealing so far with conditions before the COVID-19 virus enters the scene. We shall turn in a moment to the collapse of personal wellness due to the onslaught of the virus. One would have wished that the nations of the world would have been able to meet the new set of challenges with a good defensive position in place. Alas, our numerical results point to the opposite.

A thorough discussion and analysis of unseen lacunae of poverty in modern nations would require the insights and expertise of the trained social scientist and historian. Here we have to limit ourselves to a few cursory remarks. First, we find reason to believe that these lacunae in personal well-being—if they do indeed exist—are a fairly recent phenomenon, associated with deep- going social, economic, and even cultural structural changes occurring in the new millennium.

Next, a quick look at the seven indicators of personal wellness:

100 − OBESITY provides a snapshot of the physical health of an individual. The measure can be improved by physical exercise and balanced nutrition. An increasing degree of obesity in children is being observed in many countries. The modern mother often returns from her work in the evening, collects her children in the child care center, picks up some factory-processed ready-made dinner in the neighborhood store, and serves it in front of the home TV set. Result: obesity.

On a deeper level, obesity clearly is associated with the migration of people from the countryside to the city, the rapid growth of processed food in supermarkets, and the proliferation of hamburger chains.

Among all OECD nations, the US and New Zealand obtained the lowest 100 − OBESITY ratings (less than 70), which no doubt contributed to these two being rated as non-Pareto. Turkey did not do any better, but rated Pareto anyhow—it found compensation along other dimensions of personal wellness.

SANITA is a measure of the availability of both fresh running water and water toilets, key determinants of sanitation in the home.These features are typically more common in the city than in the countryside, and more common in more recently built homes. There is an obvious causal link between poor sanitation and the spread of infectious diseases.

Among the deficit countries, Ireland and Lithuania share poorer SANITA results than the others while Australia, Israel, and New Zealand scored top marks.

The huge waves of migrants arriving to Europe over the Mediterranean or to the US from Latin America in recent years were typically settled first in make-shift shelters and later in quickly made-available already existing housing with poor infrastructure. Poor sanitation too often goes together with crowded living conditions.

There is also the matter of falling groundwater levels in parts of the world. Many areas in California struggle with polluted groundwater and ineffective treatment and use of wastewater.

The indicator (100 − UNEMPL) measures not just current conditions in the labor market in the various OECD countries but reflects also fundamental demographical changes like young people spending more years in school or at the university, or women increasingly seeking and getting paid employment outside the home or family farm. In other cases, the employment indicator signals the decline and death of entire industrial sectors in a country, like the Rust Belt in mid-America, or on the other hand the explosive new opportunities in nonfossil energy generation.

The lowest employment rates (less than 80% of the population) were reported by Canada and Greece, the former rated as non-Pareto and the latter as Paretoan. What did Greece do that Canada was not able match? Both countries also scored below-average results for MOBILE but in addition Canada did not do well on SAFE.

Turning now to the indicator MOBILE, the various national results clearly mirrors two things: the differing speed of industrial installation of a very modern and advanced technology and the manner this new way of personal communication is received by the general public. Obviously, it takes time and considerable IT resources to extend a broadband network over large geographical areas, the lowest rate of coverage being reported by Mexico.

Communication is not just gadgets. It is changing people. To the generations of the new millennium, networking—reaching out to others—has become a way of life. Presumably it should be a large font of personal wellness. But it is not for everybody. Many people in the modern world are lonely.

The size of the households are falling rapidly in the Western world. Whereas migrant households from the Middle East, Africa, and Latin America often consist of three generations of persons (grandparents, children, and grandchildren), families in Western Europe have fewer children who leave home earlier in their lives. See *Household Size and Composition Around the World 2017*, UN department for Economic and Social Affairs.

The figures in our tables show that three countries have more than 150 broadband subscriptions per 100 inhabitants: Finland (the land of Nokia), Japan, and Poland. They are all Pareto optimal.

The indicator of poverty among old people (100 − ELDERLY) not only reflects the functioning of private and social pension arrangements in a country, but also the increasing life span of people. Rather liberally, the OECD reckons persons aged 66 years or over to be "elderly." The highest figures (more than 95% of all elderly people not being impoverished) were obtained for Czechia, Denmark, France, Iceland, Netherlands, Norway (all deficit countries), and Finland and Slovakia (these two being Pareto optimal). The lowest figure (just barely more than half of the elderly population) was recorded by Korea. Yet, Korea is Pareto optimal, reaching outstanding results for most of the other indicators.

The indicator 10 – WASTE tells us part of the story of waste control in a country, but not all. It tells us how much waste is collected by municipal authorities, but not what happens after it has been collected. In the worst case, it is simply dumped in a leaking landfill. Good alternatives are sorting for reuse, or processing into fertilizers. The lowest WASTE figures are recorded for Chile, Czechia, Latvia, Lithuania, and Poland. In agricultural areas with low population density, municipal waste collection may simply not be available.

Finally, the indicator SAFE tells us something about crime in a country, or rather the perception of common people of the prevalence of crime. A number of factors lead to increased violence in the modern world: the growing social acceptance of drug use and the worldwide grasp of the narcotics industry, the trafficking of migrants, and the presence in the inner city of criminal gangs. The highest figures of personal safety are scored by Norway and Slovenia (more than 90% of the respondents feeling safe walking alone at night). The lowest figures belong to Chile, Mexico, and Turkey.

This concludes our examination of the numerical results for our seven wellness indicators. So, is there a conclusion to be drawn, an explanation why modern Western nations are scoring below countries like Poland, Portugal, Mexico, and Estonia (the four most common peers)? We have earlier said that these deficits constitute unseen pockets of lacking well-being. Is it now possible to characterize those pockets somewhat further?

Here are a few observations about the deficit countries: More than 25% of the population in Australia, Canada, Ireland, Israel, Lithuania, New Zealand, the UK, and the US is overweight (with a body mass index of 30 or more). More than 15% of all people aged 66 or over in Australia, Israel, Lithuania, Switzerland, and the US are impoverished. Only less than 70% of the population in Australia, Belgium, Canada, Italy, Lithuania, and New Zealand feel safe enough to walk alone at night.

3.8 The COVID-19 pandemic: a tentative cognitive map of causes and effects

Clearly, the causal relationships between the pandemic and the SDGs are quite complex with the virus acting as the cause of sickness and death, and at the same time, the rate of transmission of the virus depending on key SDG achievements of a country like hygiene and resources available in the hospital and care sectors.

The cognitive map in Fig. 3.1 illustrates some of the cause–effect interrelationships between key factors. The top layer of factors shows some

Figure 3.1 Causal relations between the pandemic and the sustainable development goals, illustrated by a cognitive map (the reader is invited to draw arrows pointing from cause to effect).

relevant SDGs measuring the immediate vulnerability of a country to the COVID-19. We know that the incidence of the virus indicates substantially increased risk levels for obese people (SDG II) and for patients suffering from diabetes. The incidence increases in regions with poor access to running water (SDG VI) or in crammed settlements like refugee camps (SDG XI).

The middle layer in Fig. 3.1 lists some typical policy actions employed (with varying success) by most nations, responding to increasing numbers of positive cases and deaths. These policies impact on the SDGs in different ways, mainly through unintended negative effects on business activity but also on the welfare of private individuals. Closing the borders (ports, airports) cuts off imports and exports and hurts the tourism industry. Social distancing hurts transportation and restaurants. Closing schools, including nurseries and universities hurts ordinary families. Lockdown hurts employment. And so on.

The COVID-19 itself is featured in the bottom row of the figure. Both absolute numbers and the incidence per 100,000 individuals are important in determining government action.

So, is the virus a cause or an effect? It is a cause entering the cognitive map at the bottom row. It is also an effect, the incidence of cases and deaths listed as one of the indicators of SDG 3 (together with data on HIV, ebola and other pandemics).

3.9 Official measures of the spread of the virus

Several international observers publish regular updates on the incidence of the virus, among them the World Health Organization, John Hopkins University, Oxford University, and the private Worldometer. For our work in the present book, we have relied on the website tables provided by the WHO.

For each country, WHO lists the number of confirmed cases and the number of deaths, both daily and cumulatively since the beginning of the pandemic. Divided by the size of the population, these data can directly be converted into the number of cases and deaths per 100,000 persons.

The number of confirmed cases measures the burden placed on doctors, nurses, carers, and hospitals. If the number escalates, authorities tend to channel additional resources to these areas. When the number of deaths escalates, authorities tend to respond with measures like closing national borders, closing schools and restaurants, or even the lockdown of entire cities. In most countries, these decisions are made by top health officials rather than politicians.

The difference between the number of total confirmed cases and the number of deaths constitutes the survivors. Remembering that all indicators of the SDGs should represent a desirability, a positive goal worth striving for, we shall compute

$$\text{survival rate} = (\text{confirmed cases} - \text{deaths})/\text{confirmed cases} \tag{3.9}$$

Which factors determine the survival rate in a country? Once you have been caught with the virus and test positive, the further development of the disease depends upon your general health and upon the nursing care and/or hospital treatment that you receive. In other words, two sets of SDGs come into play: goals relating to the personal wellness of individuals and goals relating to the availability and quality of hospital care.

Collecting data for two benchmark dates, April 1, 2020 and September 1, 2020 we shall get two snapshots of the progression of the pandemic: one focusing on its early acceleration through the western world and one focusing on the hopes and disappointments of preventing a second wave. The raw data and the calculation of (Eq. 3.9) for all OECD countries are reported in Table 3.3.

The social scientist usually is at a disadvantage compared to research in physics or chemistry or bacteriology, not being able to conduct controlled experiments. On a few occasions, however, circumstances present us with

Table 3.3 COVID-19 survival rates in the OECD countries. Data downloaded from the WHO virus dashboard.

Country	Survival rate, April 1 / Sept 1	Cumulative deaths, April 1 /Sept 1	Cumulative confirmed cases, April 1 / Sept 1
Australia	0.996/0.975	20/652	4900/25,700
Austria	0.987/0.973	128/733	10,200/27,400
Belgium	0.918/0.802	1400/17,000	17,000/86,000
Canada	0.988/0.929	89/9100	7700/127,900
Chile	0.996/0.973	12/11,300	2700/411,700
Czechia	0.991/0.983	31/424	3300/24,600
Denmark	0.967/0.963	90/624	2900/17,000
Estonia	0.995/0.996	4/64	745/2400
Finland	0.988/0.959	17/335	1400/8100
France	0.932/0.887	3500/30,500	51,500/265,000
Germany	0.989/0.962	732/9300	67,400/243,600
Greece	0.962/0.974	49/266	1300/10,300
Hungary	0.950/0.899	20/615	525/6100
Iceland	0.998/0.995	2/10	1100/2100
Ireland	0.978/0.938	71/1800	3200/28,800
Israel	0.996/0.976	20/6400	5100/270,100
Italy	0.883/0.868	12,400/35,500	105,800/269,200
Japan	0.974/0.981	57/1300	2200/68,400
Korea	0.983/0.984	165/324	9900/20,200
Latvia	1/0.756	0/347	398/1400
Lithuania	0.987/0.97	7/86	533/2900
Luxembourg	0.990/0.981	23/124	2200/6700
Mexico	1/0.892	28/64,200	64,200/595,800
Netherlands	0.921/0.912	1000/6200	12,600/70,600
New Zealand	0.998/0.984	1/22	647/1400
Norway	0.994/0.975	28/264	4400/10,600
Poland	0.986/0.97	33/2000	2300/67,400
Portugal	0.978/0.969	160/1800	7400/58,000
Slovakia	1/0.992	0/33	363/3900
Slovenia	0.984/0.956	13/128	814/2900
Spain	0.977/0.938	10,900/29,500	94,400/476,400
Sweden	0.929/0.931	337/5800	4800/84,400
Switzerland	0.97/0.96	537/1700	18,200/42,100
Turkey	0.984/0.976	214/6400	13,500/270,100
UK	0.926/0.876	2500/41,500	34,000/335,300
US	0.982/0.969	2900/182,200	163,200/5,900,000

a natural experiment: the policymaker of a nation taking action and the social scientist observing the result. The famous example of using such natural experiments in economics is the joint work by Milton Friedman and Anna Schwartz, reported in *A Monetary History of the United States 1867–960,* Princeton University Press, 1963. Offering an array of case studies, they investigated the effects of monetary institutions and the conduct of monetary policy on economic activity.

Present events indeed give examples of a multitude of natural experiments as nations in desperate attempts to contain the pandemic seemed to compete in applying different and often contradictory policy measures at different times. One such widely cited experiment was conducted by the top bacteriologist in Sweden, Dr. Anders Tegnell, who decided that minimal intervention to limit the spread of the virus would be helpful to create herd immunity in the entire Swedish population. We have to leave it to future research to see whether this was a realistic goal. As seen in Table 3.3, the survival rate in Sweden actually improved slightly from April to September.

Four more nations report an increasing survival rate: Estonia, Greece, Japan, and Korea. The two latter nations had already earlier learned how to protect themselves against pandemics: the severe acute respiratory syndrome that hit Japan in 2002–2003 and the Middle East respiratory syndrome that hit Korea in 2015.

The table shows that the four countries that suffered the highest death rates (cumulative number of deaths divided by cumulated number of cases) in the beginning of the year, Italy, Belgium, the Netherlands, and the UK, were hit even worse during the fall.

3.10 A numerical illustration: expanded frontier calculations incorporating the virus survival rate as a sustainable development goal indicator

Our intention is now to revisit the ranking of the SDG performance of the OECD nations, this time expanding the dashboard of metrics to include the COVID-19 survival rates listed in Table 3.3. The limited scope of such an inquiry needs to be kept in mind:

- It means that we for the moment decide to treat the incidence of the virus as an effect variable, rather than a cause. Actually, provisions for such a treatment have already been made by the UN in their list of indicators of SDG III featuring a number of deadly diseases like cancer, diabetes, HIV, and tuberculosis.

- Unfortunately it also means that we still have to work with pre-2020 data for all goal indicators, since later data will not be available for some time. Thus, we have already at the outset built into the calculations the presumption that any causal explanation of the incidence of the virus will be traced back to historical wellness data.
- Fitting the envelope to the expanded list of goals, we obviously do not know in advance how the earlier ranking of the wellness of nations will be affected. A country formerly ranked as located on the effectiveness frontier may now—because of a poor COVID-19 survival rate—fall below the frontier. A country formerly ranked as suboptimal may now—because of an excellent survival rate—turn out to belong to the Pareto elite of nations.

For the mathematical work, we again solve the earlier OECD programming model assuming constant returns to scale. The only difference is that we this time have data for eight SDG indicators rather than seven. First using the survival data for September 1, 2020, solution yielded the following list of 16 OECD nations maximizing their policy effectiveness ratio:

Chile, Estonia, Finland, Greece, Hungary, Japan, Korea, Latvia, Luxembourg, Mexico, Poland, Portugal, Slovenia, Slovakia, Spain, Turkey. (3.10)

Each one of these 16 nations exhibit neither policy input slack nor goal slack. They are all Pareto optimal.

Comparing with the earlier list (Eq. 3.2), we see that the two lists are identical. In other words: Every OECD country that earlier—before the arrival of the COVID-19—had been rated as Pareto optimal, still on September 1, 2020 belonged to that elite. Even Spain, which by that date had registered close to half a million cases and close to 30,000 deaths still was able to maintain its good average score, helped by outstanding results on hygiene and personal safety (see the entries for SANITA and SAFE in Table 3.1) Bombing on one single SDG does not necessarily destroy your average.

Here is the list of 20 suboptimal countries, ranked in the order of falling policy effectiveness ratios:

Lithuania 0.996, Czechia 0.995, Israel 0.989, Ireland 0.987, Denmark 0.983, France 0.980, New Zealand 0.976, Iceland 0.975, Italy 0.975, UK 0.973, Netherlands 0.969, Austria 0.967, Canada 0.965, Sweden 0.962, Norway 0.960, Belgium 0.958, Germany 0.958, Switzerland 0.956, Australia 0.955, US 0.943. (3.11)

They all exhibit both some positive policy input slacks and some positive goal slacks.

Remembering our earlier results (Eq. 3.4), we see that most nations earlier ranked as non-Pareto still retains exactly the same position in the ranking list. There are only two exceptions. Lithuania early recorded a policy effectiveness ratio of 0.990; this time Lithuania has climbed to the top position in the list with the ratio 0.996. New Zealand earlier recorded the policy effectiveness ratio of 0.975, this time it has climbed ahead of Iceland and Italy, obtaining the effectiveness ratio 0.976.

These minor adjustments should not distract our attention from the overwhelming conclusion that needs to be drawn.

Expanding the dashboard of metrics assessing the personal wellness of nations to include the survival ratio of COVID-19, it turns out that the very same list of countries that earlier were rated as Pareto optimal (some them being quite GDP-poor, like Chile, Greece, Latvia, and Mexico) still belong to that group. And the very same of list of countries that earlier were rated as having unrecognized pockets of personal wellness deficits (some of them being among the world's richest economies, like Canada, Germany, the UK, and the US) still stay sub-Pareto, below the performance frontier.

What does this mean? It seems to indicate that the incidence of the COVID-19 in some statistical sense could have been predicted already before the virus struck, hitting the very countries that already harbored documented wellness deficits. The lowest survival rates go together with obesity, bad hygiene, loneliness, lacking care of the elderly, a skewed income distribution, poor waste collection, and low personal safety.

It remains to look at how nations coped the COVID-19 at an earlier stage of the pandemic, on April 1, 2020. Running the optimization program again, this time using the survival data for that date, the following nations turned out to be located on the SDG performance frontier:

Chile, Estonia, Finland, Greece, Hungary, Japan, Korea, Latvia, Luxembourg, Mexico, Poland, Portugal, Slovenia, Slovakia, Spain, Turkey. (3.12)

Again, exactly the same list (Eq. 3.2) that we obtained without the COVID-19 data. The list for the subfrontier countries and their effectiveness ratings is also the same. The small adjustments in the effectiveness ratings for Lithuania and for New Zealand observed on September 1 have not yet set in:

Lithuania 0.990, Czechia 0.995, Israel 0.989, Ireland 0.987, Denmark 0.983, France 0.980, New Zealand 0.975, Iceland 0.975, Italy 0.975, UK 0.973, Netherlands 0.969, Austria 0.967, Canada 0.965, Sweden 0.962, Norway 0.960, Belgium 0.958, Germany 0.958, Switzerland 0.956, Australia 0.955, US 0.943. (3.13)

If these conclusions hold up, the message to politically responsible bacteriologists and health officials does not seem to be good: regardless of what you do in terms of social distancing or lockdown, the incidence of the virus will stay the same as it runs its course. The damage was done even before the virus arrived.

In order to double-check the results now reported, we reran the OECD programming model fitting a variable-returns-to-scale envelope instead. Using the COVID-19 data for September 1, 2020, we obtained the following list of Pareto optimal countries

Australia, Austria, Belgium, Chile, Czechia, Denmark, Estonia, Finland, France, Greece Hungary, Iceland, Ireland, Israel, Japan, Korea, Latvia, Lithuania, Luxembourg, Mexico, Netherlands, New Zealand, Norway, Poland, Portugal, Slovenia, Slovakia, Spain, Switzerland, Turkey. (3.14)

which is identical to our earlier list (Eq. 3.6). Consequently, the list of non-Pareto countries

Canada 0.999, Germany 0.999, Italy 0.999, Sweden 0.999, UK 1.000, US 1.000 (3.15)

is identical to (Eq. 3.7). One small difference, however: this time the UK and the US both achieve a policy effectiveness rating equal to 1.000. They are optimal, but not Pareto (both exhibit positive policy input slacks and goal slacks).

Solving the variable-returns-to-scale program for April 1, 2020 exactly the same numerical results obtained.

CHAPTER FOUR

Disequilibrium and chaos

While Tinbergen in the early 1950s wrote his pathbreaking treatises on economic policy, a consolidation of the inherited economic theory occurred—the codification of neoclassical theory. A generation of young mathematical economists such as Paul Samuelson, Ken Arrow, Gerard Debreu, and Tjalling Koopmans were to formalize the building blocks of an impressive mathematical idealization of a free market economy. The economic agents in the neoclassical economy are abstract consumers and producers, suspended in time, making the rational decisions of "economic man." Markets are in equilibrium. And, Tinbergen would say, economic policymakers make optimal decisions.

While this theory dominated the economics profession as the 20th century was drawing to its close, a set of revolutionary ideas were promoted by a few dissenting voices, clamoring for greater realism in economics such as evidenced by path-dependency, disequilibrium, and chaos. Here we discuss some of these ideas: how they apply to a theory of economic and social policy, and how the presence of disequilibrium can be demonstrated empirically by envelopment analysis.

The discovery of disequilibrium and the explicit introduction of time (the "arrow of time") in real-world processes of change occurred first in physics and thermodynamics. Economists, too, eventually realized that the processes they study are inherently irreversible. The world changes and we change with it. We cannot go back to the economic conditions ruling in the 1920s or even in the 1960s. New technology is developed. Consumer preferences evolve. Tropical rainforests are razed. The stratosphere becomes polluted with greenhouse gases.

4.1 Path-dependency, lock-in, evolution, creative destruction

The Santa Fe Institute is an offshoot from the nuclear weapons laboratories and computer laboratories at nearby Los Alamos, offering an opportunity for hard-core physicists and computer jocks to mix with scholars in

Measuring Sustainable Development Goals Performance
ISBN: 978-0-323-90268-7
https://doi.org/10.1016/B978-0-323-90268-7.00004-9

"soft sciences" like economics and social sciences. The institute has become renowned for its work in the study of complex behavior in mathematical, physical, and living systems.

Discarding the standard notions of equilibrium, economists working at Santa Fe developed a new kind of economic system with "positive feedback," like real estate booms and stock market crashes. Once such systems get rolling, they keep snowballing and feed themselves, up to a point. Positive feedback arises in an economy when there are increasing returns to scale. If the average production costs of some firms fall with increasing output, the managers of these firms will find it profitable to expand their scale of operations. Outputs and market shares of individual firms will embark upon a dynamic path of growth.

Examples of positive feedback in economic and social policy can be found in most of the pressing economic and social agenda of the present day. The scramble to develop new green technology is leading to a rapid buildup of both wind and solar power. Once some particular technology, like offshore wind turbines, has been demonstrated as viable, there is a rush to develop these possibilities. The average cost of manufacturing solar cells is falling rapidly; as a result, the solar cell industry in China is growing dramatically, and companies and private persons around the world are turning to this new medium of energy generation. There is a wind turbine boom and there is a solar power boom.

Powerful mechanisms often reinforce the positive feedback among a pack of competing firms. There is synergy in the development of high technology. Corporations learn not only from their own experience but also from their competitors. Each manufacturer does not have to invent the wheel all over again. One can climb on others' shoulders, adapt existing technical designs, and take note of what seems to sell for competitors and what does not sell.

Product development is one of the most important occupations of firms in the high technology age. Sometimes product development takes the form of a discrete jump in technology, a Schumpeterian "innovation," but more commonly it is about improving the technological characteristics or the marketing appeal of an already existing product. Modern products are almost invariably multidimensional, that is, they have an entire vector of attributes. Product development enhances the attributes of an existing product or it adds new attributes.

The adoption of a novel technology sets the stage for subsequent product development and innovation. Any one given vintage of technology will eventually be superseded by even more advanced technology. Consider, for instance, the technology chain: dial telephone, touch tone telephone, cordless telephone, mobile phone, "smart" phone (with access to the Internet). The graph of the network buds as new links and new nodes are added hierarchically.

In the short run, budding in the network involves mainly growth at the top hierarchical layer of the network. The network grows at its periphery. But in the long run, as the network expands, what was once the periphery may eventually become the core of entire families of novel technologies. Choices of technology that were once marginal may define the fundamentals of the technologies of the future. As a consequence, choices that once seemed insignificant may eventually turn out be crucial. Technologies become "locked in." As an example, Santa Fe economist B. Arthur (1989) quotes the competition between steam and the internal combustion engine for the propulsion of motor vehicles at the end of the 19th century. Once the Otto engine had been chosen, the stage was set for the development of successor technologies like the Daimler engine, the spray carburetor, the honeycomb radiator, the self-starter, and so forth. The steam alternative was never pursued.

When self-reinforcing mechanisms are at play, there is little reason to expect the resulting growth path to be orderly and linear. Companies and government entities will scramble to exploit emerging opportunities. New startup companies will enter the fray. Arthur suggested that the resulting dynamic economic system can possess a multiplicity of solutions, that is, a multiplicity of alternative future courses. Many candidates for long-term self-reinforcement may exist; the accumulation of small random events early on pushes the dynamics into the orbit of one of these and thus selects a time path that the system eventually settles on.

These technological webs can undergo bursts of evolutionary creativity and massive extinction events, just like biological ecosystems. Consider the invention of the automobile, replacing an older technology, the horse. Along with the horse go the smithy, the Pony Express, the watering troughs, the stables, the people who curried horses, and so on. The whole subnetwork of technologies that depended upon the horse suddenly collapses in what economist Joseph Schumpeter once called "the gale of creative destruction." It is "creative" because in its wake whole new subnetworks of goods and services would appear, paved roads, gas stations, fast food restaurants, motels, traffic courts, traffic cops, and traffic lights. [Waldrop, 1992, interviewing W. B. Arthur.]

Evolution of new goods and services (and their eventual obsolescence) occurs not only for industrial products but also for medical, educational, and government services that are the subject to economic and social policy. New hospital wards built to accommodate the latest medical progress will eventually need to be refashioned entirely. Entire hospitals are closed down (as happened with most of the tuberculosis sanatoriums 100 years ago). New university research facilities are being built, but will ultimately need to be refitted to pursue entirely new lines of research. New commercial airports are being built and others are closed down. (When the downtown airport of Oslo closed its doors, a hectic phase of creative destruction set in, building a new high tech city on land formerly occupied by runways.) What will happen to the NASA installations when the private sector takes over all space flight?

Still keeping Tinbergen's policy optimization problem in mind, there is the added difficulty that not only are the various technologies of generating economic and social services in modern society subject to rapid development—the *political evaluation* of the urgency of these services is also in a continuous path-dependent flux. Scientific knowledge increases—medical breakthroughs occur, genetic engineering leads to new farming opportunities, seismic probes reveal the presence of untapped reserves of oil and gas, improved engineering designs lead to more energy-efficient transport. But the political will to translate such progress into government policy (the willingness to fund extensive inoculation programs in Africa, to permit the sale of genetically modified foodstuffs, to open up Arctic waters to oil and gas exploration, to impose mandatory rules on the fuel efficiency of cars sold to the general public) is often hotly debated, and sometimes changes dramatically over time. In language that Tinbergen would have understood: the overall social utility function itself shifts over time. It, too, is path-dependent.

4.2 Diffusion, self-organization, and chaos

The general idea that evolution takes the form of a succession of generations of "species," be it economic species or biological species, is as old as the concept of evolution itself. In the marketplace, a novel and attractive product will capture ("steal") market share that would otherwise have belonged to competitors selling outmoded product designs. The buyers face the opportunity of switching from an earlier product generation to a later one. The same phenomenon occurs in the biological world when dandelions and crabgrass take over a lawn. The capture occurs as a crowding out, the taking of potential living space.

The mathematics of successive generations of species was developed in the 1930s by the Italian mathematician Vito Volterra. His classical treatise is titled *Lecons sur la Theorie Mathematique de la Lutte pour la Vie (Lessons on the Mathematical Theory of the Struggle for Life).* Here one finds the origin of the modern nonlinear mathematical theory of evolution. Biological evolution occurs among plants and animals, as new species compete for living space. It is irreversible in time. Many of the laws describing the growth, decay, and interaction of biological populations can also be applied to social, economic and financial systems. Evolution is not always orderly. It can take the form of sudden invasions of new species and the extinction of others.

In 1975, the mathematician J. Yorke was working with a diffusion curve programmed on a computer, examining its development over time. Using a continuous time variable in the mathematical formulation, the standard diffusion curve appeared on the screen, rising quickly at first, and then slowly approaching its saturation level. Replacing continuous time by discrete time (cutting up the time dimension in a series of still segments), the discrete time path of the diffusion curve converges toward the continuous path. It may temporarily overshoot the saturation level, but eventually zooms in toward the equilibrium value. Changing the parameters of his numerical model slightly, Yorke made a startling discovery. The curve became jittery and "bifurcated" (it split into two). The diffusion path ended up oscillating between two alternative values, one greater saturation level and another smaller saturation level.

Bifurcation is a balancing act between order and full-blown chaos. It is a fleeting moment before the system takes the plunge into disorder. It is the sudden appearance of alternative futures, a forking point along the road of technological advance, such as two competing product designs. Changing the parameter even further, the discrete path broke up into chaos. And yet, it had been generated from a very simple and deterministic mathematical rule. Yorke's paper was published in a mathematical journal in 1975. A new science had been born.

Is chaos the same thing as randomness? Chaos can be produced by a few simple nonlinear mathematical equations. The path of the chaotic process over time is determined, point by point, by an exact mathematical formula.

It is "deterministic." Randomness, on the other hand, is the result of probability drawings, like the random throw of dice. Chaos looks random, but there is order hidden behind the apparent whimsicality.

The concepts of bifurcation and chaos extend to social and economic policy and the competitive battle between alternative "vintages" of policy. Consider education policy as formulated by Western pedagogy and the educational alternatives offered (or rather not offered) by militant Muslims, closing schools to all female students. In parts of Afghanistan, Pakistan, and many Muslim countries in Africa, the girl in the classroom is seen as an abomination. As local militia expands their control, female students are dismissed and chased home. As the legal government reasserts itself, they return. A bifurcation of two competing world-views of educational policy.

The tumultuous events in the euro market in the fall of 2011 provide an instructive example of full-blown chaos. How does a country within the euro zone formulate an "optimal" policy of managing its government debt portfolio, when the cost of issuing Treasury bonds denominated in euros is not only unknown, but swings from the attractive to the prohibitive within seconds? Some of the largest commercial banks in Western Europe were willing to enter into wildly speculative positions—at one moment confident that their own government would bail them out if they collapsed, the next moment struck with fright that the entire euro system would cave in. The disequilibrium processes that swept through the markets were not only irrational but bordered on panic.

4.3 Dissipative processes and disequilibrium

Ilya Prigogine was awarded the 1977 Nobel Prize in chemistry for his contributions to nonequilibrium thermodynamics, particularly for his theory of dissipative structures. He has emerged as the proponent of an evolutionary conception of the world, replacing the Newtonian mechanistic paradigm. He has shown that the behavior of matter under nonequilibrium conditions can be radically different from its behavior at, or near, equilibrium. This difference introduces multiple choices, self-organization, and complex dynamics.

Close to equilibrium, a description of the temporal evolution of a system can be expressed by linear equations. Far from equilibrium, one deals with nonlinear equations, which may lead to bifurcations and to the spontaneous evolution of organized states of matter and the appearance of dissipative structures. As an example of a dissipative structure, consider a pan of liquid heated from below. When the temperature is low, heat passes through the liquid by conduction. As the heat is intensified, regular convection cells appear spontaneously. The liquid boils. Energy is transferred from thermal motion to convection currents. The boiling dissipative structure is radically different from the equilibrium structure of the liquid. It can be maintained in far-from-equilibrium conditions only through a sufficient flow of energy. Yet, it represents order—order at the edge of chaos.

Dissipative structures can be found also in man-made systems such as social and economic policy. Consider the "Arab spring" of 2011—virtually every Muslim country suddenly faced waves of social discontent and unrest—as the ruling power elite was challenged by explosive demands for free expression and assembly. Why did these local "convection cells" appear at the same time, causing the entire Middle East to boil? What was the nature of the "energy" that sustained it?

In the book *Self-Organization in Non-equilibrium Systems: From Dissipative Structures to Order through Fluctuations*, Nicolis and Prigogine (1977) discuss the evolution of ecosystems such as insect societies where each species occupies a niche. The new generations of species (arriving through genetic mutations) are able to exploit the environment ever more effectively. The increasing effectiveness to exploit each resource provides a basis for understanding the notion of "the survival of the fittest" familiar from Darwin's theory.

How do these developments impact on economic theory? For one thing, they show that economists need to consider the arrow of time, irreversible processes, and the possibility of far-from-equilibrium positions. The Newtonian theory of mechanistic and timeless equilibrium theory is not dead, it just needs to be imbedded in an irreversible and evolutionary setting. Similarly, the great achievements of neoclassical economic theory are not dead, but need to be imbedded in a setting of technological and financial innovations, increasing returns to scale, path dependency, and the possibility of far-from-equilibrium processes. Economists need to construct a new theory of market formation, where the actors—producers and consumers—are imperfectly adjusted to their theoretical equilibrium. In extreme cases, they may be pursuing individual orbits of hyper-growth, or orbits of eventual extinction. Just as in a dissipative process of boiling water, the actors of the market may collectively form a far-from-equilibrium state of dynamic swarming of the individual cells but maintaining unchanging macroproperties—the edge of chaos.

The same mathematics can be used to illustrate the chaotic super-growth of high-tech companies like Google or Apple, or the super-growth of countries like China and India. These chaotic growth paths are in wild disequilibrium. But they define platforms of new order.

4.4 The arrow of time

Prigogine belongs to that rare brand of physicists and thinkers who have helped define not only what we see in the world around us but also what we are, and our role in the universe. The first steps along this road were taken by the Greek philosopher Epicurus (341−270 BC). As a follower of Democritus, he believed that everything that has ever existed and that will ever exist is made out of indestructible building blocks, irreducibly small in size and unimaginably vast in number: atoms.

In his book *The End of Certainty* (1996), Prigogine discusses the fundamental dilemma raised by Epicurus: What can be the meaning of human freedom in a deterministic world of atoms? Epicurus used atomistic theory to develop a philosophical system seeking ethical ends. If the simplest fact of existence is rational and if there exists a natural explanation for everything, then man would be liberated from superstition, even the wrath of the gods, and could instead pursue a life of seeking pleasure in simple things. Epicurus lived a frugal life in a secluded garden in Athens.

But the great physicists of times who followed him would be haunted by doubts. With natural sciences revealing the great laws of the universe, what role is there left for the human mind? Prigogine cites Einstein:

> *...a Being endowed with higher insight and more perfect intelligence, watching man and his doings, [would] smile about this illusion of his that he was acting according to his own free will.*

If you believe that the world is just one big automaton following mathematical laws, then what room is left for human reasoning, hopes, and dreams? Like Einstein, a generation of modern physicists has been filled with pessimism and Weltschmerz. Particularly, there is the Second Law of Thermodynamics which traditionally is understood to place mankind in crushing dilemma, formulated by Rudolf Clausius in the mid-19th century. He developed a theory for the working of the steam engine, stating that "heat cannot by itself pass from a colder to a hotter body" and stressed the concept of entropy (dissipation of available energy). Whenever work is done, an amount of usable energy is irrevocably lost. All matter and energy in the universe will eventually degrade into a state of inert uniformity or heat death.

And so, the world clock ticks slowly and inexorably toward the final extinction of the universe and of time itself. For some, such as Stephen W. Hawking, in his *Brief History of Time*, we are close to the end. Prigogine asks "How can we conceive of creativity or ethics in a deterministic world?" [ibid., p. 6].

More than a decade has passed since Prigogine died peacefully in Brussels. Already it is becoming clear that posterity will look upon Prigogine as one of the greatest physicist since Newton. Not only did he make some of the most fundamental discoveries of his age in chemistry, thermodynamics, and physics—but, more importantly, he has given back to modern man a belief in himself and in the future. Following in Prigogine's footsteps, scientists continue to discover new patterns on all levels of existence, from the biology of molecules to the cosmology of the universe. Prigogine transformed physics, giving it a new cultural relevance. In the arts, medicine, business, social science, and technology, professionals have acknowledged their debts to his fundamental and resounding insights. As Prigogine himself put it:

> *Mankind is at a turning point, the beginning of a new rationality.... We are observing the birth of a science that views us and our creativity as part of a fundamental trend present at all levels of nature.*
>
> ***(ibid.)***

As a young student at the Université Libre in Brussels, Prigogine hit upon the key question that was going to pursue him relentlessly during his entire scientific career: What is time? Time was not mentioned in his physics classes. Physics did not deal with time. Newton's system of planetary motion is reversible in time; it works just as well backwards as forwards. The movements of a pendulum can be repeated exactly backwards in time.

There is an "arrow of time" pointing away from the past and toward the future. To a physicist brought up into a Newtonian world of mechanistic and time-invariant laws, this is a most disturbing fact. You cannot "undiscover" the inventions of the Apple iPhone, iPod, or iPad. The company can go under, but the technology will remain. And you cannot run the huge economic boom in China backwards in time. The growth rate may slow down, but the new Chinese middle class will have their ambitions and dreams raised forever. The arrow of time points forward.

Actually, history itself is irreversible in time, starting with the moment of the big bang and reaching up to our own times. The reversible and timeless Newtonian system should not be discarded, but it needs to be amended to account for the presence of irreversible processes, opening up the possibility of evolution. Evolution occurred on our planet even before the advent of life: the slow cooling of the planet, and the formation of geological strata, rivers, and seas. The historic formation of the Nile with its tributaries and the climatic conditions created a "time clock" of alternating seasons of flooding in upper Egypt, generating an order of regular growing and harvesting seasons downstream, thus putting in place the conditions for one of the first civilizations on our planet.

The subject of evolution of plants and animals brings us to Darwinian evolution. In many ways Prigogine emulates what Darwin did in biology, formulating an evolutionary approach to physics. Extending the mathematics of chemical thermodynamics, Prigogine discusses the concept of a "niche"—a population of molecules or biological individuals that stays unabsorbed and maintained inside a larger medium: a pocket of gas or a liquid, or a biological species.

Nicolis and Prigogine write:

> *The increasing effectiveness of the exploitation of each resource provides a basis for understanding the notion of "survival of the fittest" familiar from Darwin's theory. In particular, our arguments show that a definition of the "fittest" can only be obtained through a suitable analysis of the evolution equations describing the ecosystem.*
>
> **(ibid, *p. 458*)**

Deterministic processes can be seen only in isolated, artificial systems. The natural world, on the other hand, is a world of irreversible processes, of self-organization. Prigogine perceives nature on Earth in exactly the opposite way it was viewed in classical physics. He no longer conceives of nature as a passive object. It is an active object in our lives. What we see here on Earth is just the opposite of entropy. Instead of approaching a heat death, we see successive diversification. We see evolution. Prigogine views energy dissipation as the driving force of evolution. He notes that, despite the increase in organization and complexity of living systems, there has been an acceleration of biological evolution in the course of time. Each new step increasing the functional organization has in itself the germs for further evolution.

Prigogine offers a guardedly optimistic alternative to the pessimistic view of mankind's future—entropy and all-reigning disorder spreading throughout the universe. In an interview with *OMNI* magazine in 1983, he said:

> *The classical view divided the universe between spiritual self and the physical, external world. Yet inside us we see time, activity; we experience change. This internal experience is in complete contrast with the view of the world as a timeless automaton. As we begin to discover the roots of time outside us, this duality tends to disappear. We see a convergence between the world outside and the world inside us. With the paradigm of self-organization, we see a transition from disorder to order. In the field of psychological activity, this is perhaps the main experience we have—every artistic or scientific creation implies a transition from disorder to order.*

Prigogine discusses the "narrow path" in the last chapter of *The End of Certainty* [1997]:

> *We have tried to follow … a narrow path between two conceptions that both lead to alienation: a world ruled by deterministic laws, which leaves no place for novelty, and a world ruled by a dice-playing God, where everything is absurd, acausal, and incomprehensible ….We have attempted to make this book a journey along the narrow path, and thereby illustrate the role of human creativity in science.*
>
> ***(pp. 187–88)***

CHAPTER FIVE

The founding fathers of data envelopment analysis

A. Charnes and W.W. Cooper

The idea to use mathematics and computers as a tool to solve management problems goes back to World War II, when mathematicians would calculate optimal bombing runs over enemy country, or the deployment of submarines to protect convoys. The new mathematics developed for this purpose was called "operations research." After the war, the idea was born to adapt the same techniques to some kind of "scientific" conduct of regular business operations—"management science." Two intellectual giants stood out in these efforts, young researchers at the Graduate School of Industrial Administration, Carnegie Mellon University: A. Charnes and W. W. Cooper. In pathbreaking papers, they would establish management science as a reputable and impressive new field of scholarly research. Their two volumes of *Management Models and Industrial Applications of Linear Programming* (1961) summed up the essentials of a flood of some 40 papers on the subject that they published, jointly or separately, during the preceding decade.

They had originally met over lunch at Carnegie Institute of Technology in 1948 when Charnes began his postdoctoral career as an assistant professor of mathematics. The key to the extraordinary fecundity of their research was no doubt their strong bonds of friendship, a strong sense of joint purpose, and their accelerating insight into the potential of the new operations research techniques that they invented. Some sciences are more or less sterile, having been harvested over and over again. Operations research under their hands became a new exploding science.

In these efforts, Charnes and Cooper struck up a lasting friendship with George Kozmetsky, a young researcher at Harvard. Kozmetsky decided to test the theory that this new mathematics could be used to make money. He founded Teledyne (with Henry Singleton), which would become one of the largest industrial combines in the nation. The three friends (with others) started The Institute of Management Science, a society having the objective of "identifying, extending, and unifying scientific knowledge that contributed to the understanding and practice of management."

Measuring Sustainable Development Goals Performance
ISBN: 978-0-323-90268-7
https://doi.org/10.1016/B978-0-323-90268-7.00013-X

Cooper was its first (and founding) president. Kozmetsky guaranteed the financial standing for the society's new journal, *Management Science*. Charnes became its vice president, then president, and the editor of *Management Science*, the society's journal.

5.1 The early days of linear programming

Around the year 1947, John von Neumann presented in his lectures the main features of linear programming, including the simplex method; he also indicated the nature of the dual theorem. His work was continued by his students including G. Dantzig. Much of this work was published in the *Cowles Commission Monograph no. 13*, 1951. It should be noted that at this early date, no practical application of linear programming had ever been reported in the literature. There was no clear understanding of the interpretation of the dual problem. Only in the summer of 1951, at a meeting of the American Mathematical Society, Charnes and Greenberg presented a practical example where a straightforward interpretation could be accorded to both the direct and the dual variables.

Early attempts at practical application of linear programming had invariably stranded because the simplex calculations ran into indeterminacy and degeneracy. At the time nobody knew how to deal with this. The problem of degeneracy was finally solved by Charnes in his paper "Optimality and Degeneracy in Linear Programming," (1952). His solution introduced so-called non-Archimedean field extensions into the mathematical programming literature—a mathematical strategem to which he would repeatedly return in his future research.

When Charnes submitted his paper to *Econometrica*, the editor W. B. Simpson sent it on to George Dantzig for review. This was not a good choice. It was difficult for Dantzig to write the referee report in an impartial manner since he was known to have strong negative feelings about Charnes. Upon the advice of L. Hurwicz, Simpson therefore passed the refereeing task on to N. Georgescu-Roegen, and the paper was immediately accepted for publication. Dantzig responded by submitting a "Note on Charnes's paper on optimality and degeneracy in linear programming" to the journal. Simpson sent it on to Charnes so that he could answer in the same issue of the journal—a common practice. Charnes did indeed write an answer "Rejoinder to G. B. Dantzig: a comparison of perturbation methods" pointing out mathematical errors in Dantzig's opus. Dantzig then withdrew his note; it was never published, nor was the Charnes Rejoinder ever published.

With the degeneracy problem satisfactorily dealt with, it was finally possible to turn to a real-life application of linear programming: the blending of aviation gasoline. It was a classic joint effort, Charnes having developed the math, Cooper doing the calculations on an old-fashioned hand-operated calculator, and B. Mellon of Gulf Oil Company supplying the specific industry knowledge. This pathbreaking work appeared in the April 1952 issue of *Econometrica* (Charnes, Cooper, and Mellon). In one stroke, the ideas of J. von Neumann had been converted into a practical and easy-to-use tool.

A series of other practical applications followed. These included assignments for General Electric (leading to the development of goal programming) and for the Standard Oil Company of New Jersey (chance-constrained programming).

In retrospect, it is not easy to understand how the most famous research team in operations research in the world actually operated. During some of their most productive years (1957—1968), they lived 500 miles apart—with Charnes at the Technological Institute, Northwestern University on the shores of Lake Michigan, and Cooper never leaving their common alma mater Carnegie Institute of Technology (rechristened Carnegie Mellon in 1965) in the steel town of Pittsburgh. Communication was not made easier by the fact that Charnes preferred to travel by train in the 1960s. (He claimed that he would get a lot of work done on the train, but the truth was that he was afraid of airplane accidents, at the time not uncommon).

Charnes was a magnetic teacher. His graduate seminar at the Technological Institute was jammed with people, many of them coming straight from the department of mathematics. Two students would be assigned to write an abstract of each class, type it, and distribute it in mimeographed form for the next class meeting. The task was sought after, but not easy, because the material that Charnes would cover would sometimes be made up on the spot, only to appear in a new journal article some months later, perhaps even featuring the student as a coauthor.

5.2 Goal programming

The initial developments of goal programming were presented in Charnes, Cooper, and Ferguson (1955), and involved setting goals for executive compensation. Typical applications to public or social management have dealt with resource allocation in a marine environmental protection program, and budgetary decisions in a consumer protection regulatory agency. (For reference to these works and many others, see the survey article Charnes and Cooper (1977).)

In 1968, Bill Cooper and Otto Davis had prepared a proposal to the Richard King Mellon Foundation to create a School of Urban and Public Affairs (today renamed as the Heinz School of Public Policy and Management). The proposal found favor with the Mellon family who donated 10 million dollars for the purpose. Cooper became the first dean. Directing the educational program and the research carried out at the new school, Cooper became one of the early pioneers of the branch of economics that today is known as "environmental economics."

It is easy to explain how goal programming can account for environmental and social considerations. Suppose that a policymaker feels that the exploitation of some resource X is excessive and detrimental in the long run, such as the deforestation of wildlife tracts and tropical rain forests, eventually leading to erosion of soil and climate change. The employment of other resources may be observed as lacking, such as the use of unskilled labor, workers with handicaps, etc. In short, assume that it is possible to encapsulate the various considerations regarding the use of the resource into a single numerical goal g and write:

$$X = g + g^{+} - g^{-}$$

Deviations from the goal are permitted, both in the positive direction (an overfulfillment or excess is here denoted by g^{+}) and in the negative direction (an underfulfillment or deficit being denoted by g^{-}). Deviations from the goal are associated with penalties. For instance, in the case of supply of ground water, there will be some large penalty to be paid for overwithdrawal from the aquifer. No penalty will be levied for withdrawals of ground water that fall short of the goal figure. Let the unit penalties be M and N, respectively, and total penalties are then:

$$Mg^{+} + Ng^{-}$$

The larger the penalty that is associated with deviation from a goal, the more likely it is that the goal will be satisfied. Whether penalties are charged in cash or are suffered as a psychological inconvenience, in the goal programming formulation they are, in all circumstances, counted as costs of doing business—like costs for raw materials, power, water, etc.

A key question in all goal programming is this: Who will enforce the optimal solution? Clearly, some mechanism of enforcement is needed, otherwise the solution will revert to the free market solution. Few producers will voluntarily adjust their use of inputs unless they are induced to do so by being forced to pay goal penalties, or by being paid goal subsidies, directly or indirectly. There are several answers to this question. One possibility is social and attitudinal change: some producers may become aware of the urgency of environmental factors; they will then act as if bonuses and penalties are assessed for goal deviations. But no bonuses would actually be paid and no penalties collected in dollars and cents. Another possibility is that local or central government actually pays the bonuses and collects the penalties in cash. Examples of the latter are provided by the US steel industry, which since the mid-1960s has had to comply with environmental guidelines, and pay cash fines for excessive pollution of air and/or water.

A third possibility is that local or central government computes the dual variable (shadow price) of the goal constraint, as shown above, and introduces a system of unit *subsidies* or *excise taxes* connected with the use of resources. The possibility of such subsidies and excise taxes was discussed in 1915 by A. C. Pigou in his book *The Economics of Welfare*. The use of Pigou taxes has become widespread in the modern economy, for instance, as excise taxes on cigarettes, liquor, and gasoline.

Speaking in terms of modern welfare analysis, the purpose of the Pigou tax is to *internalize* the negative external economies associated with the use of the good in question. But a Pigou tax is not the only way this can be done. Sometimes it is possible to define property rights to the environmental factors which are being damaged, and to let the owner of these property rights charge a market price for the use of his property. For instance, the owner of a large wilderness tract may charge cash for hunting permits, or the owner of a river may charge a fee for permitting an upstream factory to release toxic waste into his waters.

The beauty of Charnes-Cooper goal programming is that it provides the analyst with a computing tool that not only calculates the optimal possibilities of satisfying as closely as possible the environmental targets but also in one stroke via the corresponding dual program supplies the optimal Pigou taxes (and/or subsidies).

5.3 Three friends closing ranks

Cooper had become something of a national celebrity and other universities eyed his success at Carnegie Mellon with jealousy. Harvard University made him an offer that he "just could not refuse." He spent five years at Harvard (1975–1980), but he never felt at home there.

George Kozmetsky, who still owned a major part of the industrial conglomerate Teledyne, had left running the company to his partner Singleton so that he could return to academia. Having demonstrated to himself and the world that "scientific management" could be harnessed to build an industrial empire, Kozmetsky now set himself the task of building a world-leading business research environment. To this end, he accepted an offer to become dean of the School of Business Administration at the University of Texas. One of his first moves as dean in 1968 was to sign up Abe Charnes as a faculty member, heading a new "center for cybernetic studies."

In 1977, a large symposium was organized by Charnes's many students all over the world to honor him on his 60th birthday; the proceedings were held at the University of Texas at Austin (Kortanek and Fiacco, editors, 1980). Charnes had the ability to attract outstanding students that remained warmly attached to him over the years. On the same occasion, the participants had a chance to admire the gleaming new facilities of the business school, now rapidly climbing to national prominence under the leadership of Dean Kozmetsky.

Among the various arms of the business school was a new kind of venture, The Institute for Constructive Capitalism (the IC2 Institute) that Kozmetsky had started to promote research in technology commercialization.

In the fall of 1983, the University of Texas at Austin celebrated the centennial of its foundation. To mark the occasion, the university conducted a spectacular funding drive, matching every dollar raised with a dollar from university funds. One man who eagerly responded to this opportunity was George Kozmetsky, at the time listed as one of the nation's 400 wealthiest individuals. Through his investment vehicle the RGK Foundation, he now established 16 new endowed "Centennial Fellowships" tied to the IC2 Institute. The institute itself (rechristened the Institute for Innovation, Creativity, and Capital) became organized as an independent research unit of the university.[1]

Kozmetsky realized the unique opportunity at hand—he could bring Charnes and Cooper together again on one single campus. Here was the most famous research team within the field of operations research who, except for two short years, had lived their careers separated from each other.

[1] The roster of IC2 Centennial fellows came to include some well-known academics, such as national security adviser Walt Rostow (economic history) and Nobel laureate Ilya Prigogine (chemistry). It also included the author of these lines, Sten Thore, named Gregory Kozmetsky Centennial fellow in honor of George's son.

Bill agreed to join Kozmetsky in Austin as a professor in the business school and as a centennial fellow at the IC2 Institute. The three friends were together again. The moment was precipitous. Abe and Bill were at the zenith of their creative powers, and a well-funded research organization was waiting to develop their work.

5.4 Data envelopment analysis

One of Bill's students at Carnegie Mellon was E. Rhodes, who for his dissertation research was evaluating the program Follow-Through—an educational program for disadvantaged children undertaken in US public schools with federal government support. The program recorded the performance of schools in terms of outputs such as student scores in reading and math, and included an index to measure a child's self-esteem. There were also statistics on numbers of teachers, a counseling index, and more. What kind of social or educational theory could be used to analyze such a bewildering array of data? Yet for a long time Bill had talked of tackling precisely such a complex problem. As it turned out, this research was to open the door to a new world of understanding of economic theory and policymaking.

The Charnes, Cooper, Rhodes paper on project Follow-Through was published in 1978. Data envelopment analysis (DEA) had been born. The novel idea of DEA is to discard the standard assumption of optimality that pervades most of economic theory. Some school districts are manifestly more efficient than others in implementing the Follow-Through program. So, instead of forcing assumptions on the analysis which are not warranted, let us accept that most of the schools are not located on an optimality "frontier" (or "envelope"). Only a few of the schools make up the frontier. The others are falling behind. Those that fall behind are ranked. That is, the calculations yield a diagnostic for each school: its deficit performance is measured.

As always, the PhD students of Abe and Bill were the first researchers to get into the act. R.D. Banker had been Bill's student at Harvard; his research prompted the development of a slightly different DEA model, permitting variable rates of return of the output variables (rather than the constant returns to scale that were subsumed in the Follow-Through program study). The BCC-model (Banker, Charnes, and Cooper) of DEA was published in 1984, to be distinguished from the CCR-model of Charnes, Cooper, and Rhodes (1978).

The numerical calculations involved in DEA can present subtleties (handling those non-Archimedean field extensions that are the tell-tale sign of Charnes's insights—see Section 5.1 above). Hence, they are dependent on access to rather sophisticated software. At the Center for Cybernetic Studies that Abe directed, DEA calculations were carried out using the software package that his students had collectively assembled. Things became easier later with the general algebraic modeling code GAMS. Today, specialized commercial program packages are readily available.

With these three building blocks in place (the CCR-model, the BCC-model, and numerical codes), the stage was set for DEA to conquer the world. A major milestone was the conference on New Uses of DEA in Management and Public Policy held at the IC2 Institute in September 1989. The resulting publication *Data envelopment analysis, Theory, Methodology and Applications*, by Charnes, Cooper, Lewin, and Seiford bears the dedication, "… to George Kozmetsky for his early championing of DEA and for his lifelong commitment to nurturing creativity." The publication of the book was delayed for a couple of years and appeared in 1994. By this time, the book recorded 472 published articles and dissertations related to DEA. Unfortunately, Abe Charnes never saw the book; he died in 1992.

Even after Abe's death, Bill remained in the forefront of DEA research, popularizing envelopment analysis to ever new fields of application and coauthoring several influential textbooks [Cooper et al., 2004a; Cooper et al., 2006]. In 2003, George Kozmetsky also died. But while his old-time friends were no longer with him, Bill's life did not change much. He still went to his office at the business school every day, approaching his own centennial.

In his last published paper (Cooper, Thore and Tarverdyan, 2010), using a logarithmic version DEA, a new interpretation of the fractional programming problem of DEA emerged: maximizing the utility/cost ratio for a single unit to be ranked, subject to the condition that this ratio remain feasible for every other unit. This observation opened the door for the research documented in the present book.

Epilogue

The purpose of the present investigation has been to document statistically how nations succeed or fail in reaching the UN sustainable development goals (SDGs). There were some surprises along the way and the purpose of these concluding notes is to try to assess the broader implications of our findings.

Unless you believe that every nation always behaves optimally, so that the recorded SDG achievements are the best that could have been achieved, one has to face the fact that performances at any given time differ and that some nations in one sense or another perform better than others. Then the achievements of nations can be compared. Even more: it should be possible to rank them, given some ranking criterion.

Unfortunately, the standard criterion of a performance index is of no help. To form the index, each SDG achievement suitably measured would have to be multiplied by its relative weight. The weights would reflect the relative priorities each nation attaches to the SDGs. They obviously differ from nation to nation. Certainly the weights cannot be assumed *a priori*. They need to be calculated. How?

The multidimensional technique demonstrated in this book employs the idea of determining the frontier or *envelope* of a scatter of observations. It identifies the nations that are located at the frontier, and those behind the frontier. Nations at the frontier are more successful than those falling behind. An easy and obvious ranking principle then emerges: the frontier nations are all pronounced "best" and the subfrontier nations are classified according to their frontier deficits. No single winner. But a number of subperforming nations.

So, how do we go about determining the SDG frontier statistically? There is a wealth of data available, collected by the UN and its agencies. For the calculations reported here, we have drawn on data assembled by the OECD in Paris. Actually, there is an *embarras des richesses* of data. There are 17 main categories of SDGs, each one characterized by an array of official indicators. Each UN member is required to submit annual data on 169 indicators.

Even if it were possible to process all these data into one single performance index, it would not tell us much, being composed by a jumble of disparate goals like protecting wildlife, limiting the CO_2 footprint, taking care of the elderly, waste collection, and vaccinations against the COVID-19. Instead, select a category of related goals of some particular interest, comparing the achievements of group of nations (not necessarily every UN member but rather some meaningful geographical subset of nations) in terms of these goals only. In other words, consider the wealth of available data as a *dashboard of metrics* for the conduct of economic policy, different metrics to be consulted for different policy tasks.

Once the metrics have been chosen, the statistical analysis should be straight-forward: determine the frontier and rank each nation relative to the frontier performance.

1. On the shortcomings of mainstream economic theory

Alas, right at the very start of the analysis, we run into conflict with mainstream economic theory. Nations identified as frontier strongholds (corner points spanning the envelope) pose no problems. They are in equilibrium—good old-fashioned Pareto equilibrium. But what shall we make of the sub-frontier nations? They represent suboptimal performance.

We got at first glimpse of the economic realities behind disequilibrium already in Chapter 1. Using the single explanatory variable of GDP and solving for the maximal SDG achievements of the 25 EU nations, we found that nine countries fell short of Pareto optimality. The list included Austria, Belgium, Germany, Ireland, Luxembourg, the UK, all rich countries with GDPs exceeding 40,000 dollars per capita. And yet they fell short of the UN goals in terms of average life span, numbers of restless youth (NEET), income inequality, leaking or damp family homes (LEAK), CO_2 footprint, and biodiversity. Why? Certainly it cannot be because of lack of money or resources, nor of lack of broad political support. Has economic growth in modern nations become a national juggernaut, rolling on and navigating according to its own inscrutable designs, crushing under its wheels all attempts to steer it in a more sensible direction?

The dismal results in terms of LEAK are particularly instructive. It does not take high technology to repair a private dwelling, to climb a roof and to replace broken tiles. What it takes, on a national level, is to support a a program of infrastructure. True economic growth requires the upgrading of all national infrastructure, from roads and railways to private homes to care for the elderly.

The lesson for the economist seems to be this: Market equilibrium and rational decision-making is fine, but to understand the real world, we need to recognize the presence of disequilibrium and nonrational or even irrational decision-making. Furthermore it is only when we analyze those latter cases that we really begin to understand what is going on.

The calculations reported in Chapter 3 reinforce these conclusions. Here the dashboard of metrics deliberately was chosen more narrowly, focussing on subjective or individual well-being but excluding important goals like the climate and biodiversity. "Well-being" is not a technical term usually recognized by economists but has recently been promoted by several authors and organizations tied to Joseph E. Stiglitz. For our present purposes, we decided on a rather wide definition of the concept, describing it in terms of seven SDG indicators: obesity, availability of basic sanitation services, the unemployment rate, human contact with other people, care of the elderly, household waste collection, and personal safety at night. Separately, we also added the survival rate of the COVID-19 pandemic on two occasions, early exposure (April 1, 2020) and at the beginning of the second wave (September 1, 2020).

This time we employed three macroeconomic variables as drivers: GDP, expenditure on research and development, and government spending on health and education. Also, the number of participating nations was greater, including all 36 member states of the OECD.

Determining the same kind of frontier as in Chapter 1 (the so-called upper convex hull), we were pleased to find that 30 nations came out as Pareto optimal before the COVID-19 virus struck. There were just six subfrontier nations this time: Canada, Germany, Italy, Sweden, the UK, and the US. But again, it is from disequilibrium that we learn what actually happened.

The truth is that a few of the richest and most innovative and most well-educated nations in the world fell short of some quite basic criteria of individual wellness. Glaring deficits occurred in particular in terms of obesity, care of the elderly, waste collection, and personal safety at night. More than every fifth person aged 66 or over in the US lives in poverty. More than a third of US citizens are obese. Less than 60% of the population in Canada feels safe walking alone at night (all data collected by the OECD).

Economic textbooks start out with a discussion of the individual household and so-called consumer demand, postulating that consumers maximize "utility." Later chapters outline the nature of production, supposedly put on the market to meet consumer demand. And yet, in some of the largest and mightiest economic countries in the world consumers apparently do not maximize anything like their true individual preferences, nor do the markets deliver what they are lacking.

We hasten to add that these thoughts are not new. In a number of publications, the OECD has investigated the concept of individual wellness and how it can be measured (See e.g. *How's Life? 2020: Measuring Well-Being*, OECD lLibrary, OECD Publishing, Paris). There is a growing literature questioning the ruling assumption that the economies of the Western world are paragons of sustainable development. Typically such arguments are built on observations of a few selected statistics. Our own contribution in this respect is to show how these matters can be investigated ranking nations in terms of their entire SDG performance.

2. On the impuissance of COVID-19 policy

Scholarly research often serves to confirm the expected. Sometimes it delivers new insights never bargained for.

When in the spring of 2020 we decided to incorporate data on COVID-19 survival rates into our list of SDG indicators, we expected countries like Italy, badly ravaged by the first wave of the virus, to entirely reorder the rankings of national SDG performance, placing Italy close to the bottom. By April 1, Italy had already suffered more than 12,000 deaths out of more than 100,000 confirmed COVID-19 cases, resulting in a survival rate of only 0.883. Little did we suspect that in retrospect Italy's standing in the rankings would not even marginally be affected by its misfortune.

How can this be possible? The answer is that Italy had already *before* the arrival of the virus been identified as subfrontier with a policy effectiveness ratio of less than one. In spite of its illustrious economist Wilfred Pareto, Italy was all the time non-Pareto. It was non-Pareto before the arrival of the virus and it stayed non-Pareto.

Again, the presence of disequilibrium serves to open our eyes. In fact, Italy was not the only country whose ranking was unaffected by the virus. The ranking of *every single OECD member state* stayed the same.

As in a Greek tragedy, the fate of every country had been sealed already before the curtain went up. Closing borders or not closing borders, lockdown or no lockdown, it was too late. The impuissance of COVID-19 policies. The *hubris* had already been put in place: the unseen and unrecognized pockets of preexisting deficits of personal wellness.

GAMS Program

```
$OFFSYMXREF OFFUELLIST;
  OPTION LIMROW = 0, LIMCOL = 0;
  OPTION DECIMALS = 3;
  SETS
   I input categories
   /GDP, RandD, HealthEd/
   J countries in order used by OECD with max 6 letters
   /US, Israel, Turkey, Lithua, Latvia, Mexico, Chile, Estoni, Japan, Spain,
    Greece, Korea, Italy,
    Portug, Canada, Austra, Luxemb, UnKing, NewZea, Poland, German,
    Hungar, Irelan, Austri,
    Belgiu, Switze, Sweden, Sloven, Slovak, Nether, France, Norway,
    Finlan, Czechi,
    Denmar, Icelan/
   R the output categories
   /GOAL2 obesity per cent of population
    GOAL6 sanita basic sanitation per cent of population
    GOAL8 unempl rate
    GOAL9 mobile subscriptions per 100 people
    GOAL11 elderly poverty rate
    GOAL12 waste kgs per day per capita
    GOAL16 safe alone at night per cent
    CORONA survival rate as of September 1 2020/ ;

   TABLE XRAW(J,I) the inputs provided by OECD
```

	GDP	RandD	HealthEd
US	54225	2.74	17.89
Israel	33132	4.25	9.52
Turkey	25135	0.88	7.00
Lithua	29524	0.85	7.74
Latvia	25064	0.44	7.61
Mexico	17331	0.50	7.21
Chile	22767	0.37	8.82
Estoni	29481	1.28	8.91

(*Continued*)

Japan	39002	3.15	12.07
Spain	34272	1.19	10.01
Greece	24574	1.01	8.16
Korea	35938	4.24	8.28
Italy	35220	1.29	10.05
Portug	27937	1.27	10.08
Canada	44018	1.61	11.69
Austra	44649	1.93	10.32
Luxemb	94278	1.24	8.34
UnKing	39753	1.69	12.04
NewZea	36086	1.28	12.04
Poland	27343	0.97	8.46
German	45229	2.94	12.96
Hungar	26778	1.21	7.99
Irelan	67335	1.18	8.41
Austri	45437	3.09	12.28
Belgiu	42659	2.49	13.28
Switze	57410	3.37	12.04
Sweden	46949	3.25	14.19
Sloven	31401	2.00	9.92
Slovak	30155	0.79	9.27
Nether	48473	2.03	12.76
France	38606	2.25	13.35
Norway	64800	2.03	14.93
Finlan	40586	2.75	12.88
Czechi	32606	1.68	9.15
Denmar	46683	2.87	14.80
Icelan	46483	2.10	12.27;

TABLE YRAW(J,R) the outputs provided by OECD and WHO

	GOAL2	**GOAL6**	**GOAL8**	**GOAL9**	**GOAL11**	**GOAL12**	**GOAL16**	**CORONA**
US	36.20	99.97	4.84	132.91	22.90	2.58	72.38	96.9
Israel	26.10	100.00	3.81	105.15	19.40	2.12	75.15	97.6
Turkey	32.10	96.37	11.14	70.52	17.00	1.77	56.84	97.6
Lithua	26.30	93.62.	6.92	79.82	25.10	1.10	61.14	97.0
Latvia	23.60	92.86	8.57	116.12	32.70	1.03	61.66	75.6
Mexico	28.90	89.22.	3.37	62.79	25.60	1.24	39.67	89.2
Chile	28.00	99.89	7.01	88.25	16.30	1.08	47.39	97.3
Estoni	21.20	99.61	5.59	133.35	35.70	1.47	74.89	99.6

Japan	4.30	100.00	2.44	180.95	19.60	1.71	75.73	98.1
Spain	23.80	99.90	14.55	94.11	9.40	2.13	84.05	93.8
Greece	24.90	98.96	21.07	63.40	7.80	2.00	57.49	97.4
Korea	4.70	99.89	3.73	112.77	45.70	1.24	72.89	98.4
Italy	19.90	99.28	10.57	87.91	10.30	2.23	63.71	86.8
Portug	20.80	99.44	7.35	68.88	9.50	2.21	75.79	96.9
Canada	29.40	98.50	5.87	72.17	10.50	2.33	82.14	92.9
Austra	29.00	100.00	5.47	134.88	23.20	2.23	64.92	97.5
Luxemb	22.60	97.61	5.36	88.10	7.70	2.31	83.53	98.1
UnKing	27.80	99.11	4.00	88.06	14.20	1.79	79.33	87.6
NewZea	30.80	100.00	5.04	101.60	10.60	3.68	69.09	98.4
Poland	23.10	98.13	4.36	154.12	9.30	0.88	66.12	97.0
German	22.30	99.22	3.72	79.78	9.60	2.11	74.46	96.2
Hungar	26.40	97.99	3.65	63.19	8.60	1.92	63.55	89.9
Irelan	25.30	92.18	5.98	101.97	6.40	3.58	77.76	93.8
Austri	20.10	99.97	5.39	88.12	8.70	2.40	83.64	97.3
Belgiu	22.10	99.49	6.46	75.15	8.20	1.33	64.77	80.2
Switze	19.50	99.89	4.78	98.56	19.50	2.61	89.24	96.0
Sweden	20.60	99.30.	6.32	122.52	11.00	1.61	77.10	93.1
Sloven	20.20	99.11	6.22	70.03	12.30	1.21	90.43	95.6
Slovak	20.50	98.93	7.27	82.64	4.30	1.37	66.70	99.2
Nether	20.40	97.73	3.79	90.77	3.10	2.12	84.61	91.2
France	21.60	98.65.	8.81	87.48	3.40	1.92	73.48	88.7
Norway	23.10	98.06	3.89	111.67	4.40	2.80	93.01	97.5
Finlan	22.20	99.45	8.25	153.76	5.00	2.13	87.57	95.9
Czechi	26.00	99.14	2.07	81.99	4.50	1.10	72.39	98.3
Denmar	19.70	99.60.	5.36	129.01	3.10	2.34	87.06	96.3
Icelan	21.90	98.78	2.84	113.31	2.80	1.56	83.98	99.5;

```
PARAMETER XNEW(J,I) resource inputs greater than 1;
  XNEW(J,"GDP") = XRAW(J,"GDP");
  XNEW(J,"RandD") = 10 * XRAW(J,"RandD");
  XNEW(J,"HealthEd") = XRAW(J,"HealthEd");

PARAMETER YNEW(J,R) desirable SDGs;
  YNEW(J,"GOAL2") = 100 - YRAW(J,"GOAL2") ;
  YNEW(J,"GOAL6") = YRAW(J,"GOAL6") ;
  YNEW(J,"GOAL8") = 100- YRAW(J,"GOAL8") ;
  YNEW(J,"GOAL9") = YRAW(J,"GOAL9") ;
  YNEW(J,"GOAL11") = 100- YRAW(J,"GOAL11") ;
  YNEW(J,"GOAL12") = 10 - YRAW(J,"GOAL12") ;
```

```
  YNEW(J,"GOAL16") = YRAW(J,"GOAL16") ;
  YNEW(J,"CORONA") = YRAW(J,"CORONA") ;

PARAMETER X(J,I) logs of resource inputs;
  X(J,I) = LOG(XNEW(J,I) );
PARAMETER Y(J,R) logs of desirable SDGs;
  Y(J,R) = LOG(YNEW(J,R));

PARAMETER
       Y0(R) goals of country currently rated
       X0(I) resource inputs of country currently rated ;
  Y0(R) = Y("US",R);
  X0(I) = X("US",I);
  SCALAR EPSILON infinitesimally small /0.000001/ ;

VARIABLES
  OBJ policy effectiveness ratio
  PSI optimal rate of inflation of goals to project upon frontier
POSITIVE VARIABLES
  EFFX(I) best practice GDP
  EFFY(R) best practice goal achievements
  LAMBDA(J) for each country j the weights of its peers
  GOALSLACK(R)
  INPUTSLACK(I);

EQUATIONS
  DEFOBJ    defined as sum of psi and all slacks
  DEFEFFX(I) written as a linear composite of all country inputs
  DEFEFFY(R) written as linear composite of all goal achievements
  INBAL(I)
  OUTBAL(R)
  CONVEX;

  DEFOBJ..       OBJ =E= PSI + EPSILON*(SUM(I,GDPSLACK(I))
                 +SUM(R,GOALSLACK(R)));
  DEFEFFX(I)..   EFFX(I)    =E= SUM(J,LAMBDA(J) * X(J,I));
  DEFEFFY(R)..   EFFY(R)    =E= SUM(J,LAMBDA(J) * Y(J,R));
  INBAL(I)..     EFFX(I) + INPUTSLACK(I)          =E= X0(I);
  OUTBAL(R)..    PSI * Y0(R) + GOALSLACK(R) =E= EFFY(R);
  CONVEX..       SUM(J,LAMBDA(J)).                =E= 1;

MODEL BCC/DEFOBJ, DEFEFFX, DEFEFFY, INBAL,OUTBAL,
CONVEX/;
```

```
MODEL CCR/DEFOBJ, DEFEFFX, DEFEFFY, INBAL,OUTBAL/;
  ALIAS (J,J1);
PARAMETER
  SLACKTABLE(J) both input and goal slack for each country
  RESULT(J1,*)
  RESULTX(J1,*) input results returned to original scales
  RESULTY(J1,*) goal results returned to original scales
  LAMBDATABLE(J1,J) country list of goal weights in forming best
  practice
  ;

LOOP(J1,
  X0(I) = X(J1,I);
  Y0(R) = Y(J1,R) ;
  EFFX.L(I) = X0(I);
  EFFY.L(R) = Y0(R);
  SOLVE CCR USING LP MAXIMIZING OBJ;

   YVIEW(J1,R)          = YNEW(J1,R);
   RESULT(J1,"PSI")      = PSI.L;
   RESULT(J1,"THETA")    = 1/ PSI.L;
   RESULT(1,"GOALSLACK") = SUM(R, GOALSLACK.L(R));
   RESULT(J1,"INPUTSLACK") = SUM(I, GDPSLACK.L(I));

   RESULTX(J1,I)        = EXP(EFFX.L(I));

   RESULTY(J1,"GOAL2")     = 100 - EXP(EFFY.L("GOAL2"));
   RESULTY(J1,"GOAL6")     = EXP(EFFY.L("GOAL6"));
   RESULTY(J1,"GOAL8")     = 100 - EXP(EFFY.L("GOAL8"));
   RESULTY(J1,"GOAL9")     = EXP(EFFY.L("GOAL9"));
   RESULTYJ1,"GOAL11")     = 100 - EXP(EFFY.L("GOAL11"));
   RESULTY(J1,"GOAL12")    = 10 - EXP(EFFY.L("GOAL12"));
   RESULTY(J1,"GOAL16")    = EXP(EFFY.L("GOAL16"));
   RESULTY(J1,"CORONA")    = EXP(EFFY.L("CORONA"));

   LAMBDATABLE(J1,J) = LAMBDA.L(J);
);

 DISPLAY "Results of CCR model", RESULT, RESULTX,
 RESULTY, LAMBDATABLE ;
```

PART II A GENEVA CONSENSUS

BY RUZANNA TARVERDYAN

CHAPTER SIX

Beyond Gross Domestic Product

A country's economic growth may be defined as a long-term rise in capacity to supply increasingly diverse economic goods to its population, this growing capacity based on advancing technology, and the institutional and ideological adjustments that it demands. All three components of the definition are important.

(Simon Kuznets, *Nobel Memorial Lecture, 1971*)

6.1 Recognizing the limitations of Gross Domestic Product

Since the inception of macroeconomic theory in the 1930s, Gross Domestic Product (GDP) has played the role of one of the most important statistical indicators of societal performance in national accounts. Even while developing the concept and definition of GDP, in a 1934 report to the US Congress, Simon Kuznets cautioned that it did not measure "welfare." Again, in his 1971 Nobel memorial lecture, he reiterated that "the welfare of a nation can scarcely be inferred from a measurement of national income." Yet, while the limitations of GDP were quickly recognized, it has been widely used as a measure of national welfare in proxy for other nonmeasured phenomena reflecting overall societal progress. Thus, the challenge to measure societal progress beyond economic and financial indicators such as the GDP has been at the center of various national and international studies for many years.

Kuznets listed the following six characteristics of modern economic growth based on conventional measures of national product and its components, population, labor force, *etc.*

Aggregate growth rates

- The high rates of growth of per capita output and of population in developed countries—both being large multiples of the previous rates observable in these countries and of those in the rest of the world.
- The rate of rise in productivity, i.e., of output per unit of all inputs, is high, even when we include among inputs other factors in addition to labor, the major productive factor.

Measuring Sustainable Development Goals Performance
ISBN: 978-0-323-90268-7
https://doi.org/10.1016/B978-0-323-90268-7.00001-3

Structural transformation

- Major aspects of structural change include the shift … to nonagricultural pursuits and, recently, away from industry to services; a change in the scale of productive units, and a related shift from personal enterprise to impersonal organization of economic firms, with a corresponding change in the occupational status of labor.
- The closely related and extremely important structures of society and its ideology have also changed rapidly. Urbanization and secularization come easily to mind as components of what sociologists term the process of modernization.

International spread

- The economically developed countries, by means of the increased power of technology … have the propensity to reach out to the rest of the world—thus making for one world in the sense in which this was not true in any premodern epoch.
- The spread of modern economic growth, despite its worldwide partial effects, is limited in that the economic performance in countries accounting for three-quarters of world population still falls far short of the minimum levels feasible with the potential of modern technology.

This list foreshadows a larger search for metrics with a wider coverage than GDP, which will be reviewed in a moment. Discussing structural change, Kuznets further pointed out that "the continuous disturbance of pre-existing relative positions of … economic groups is pregnant with conflict—despite the rises in absolute income or product common to all groups." He hoped for a flexible political and social framework to help accommodate rapid structural change and resolve the conflicts that it generates, while encouraging the growth-promoting groups in society. He also reflected on the social impact of modern economic growth, stating that "many of these effects are of particular interest, because they are not reflected in the current measures of economic growth."

Stakeholders in economic development acknowledge both the strengths and weaknesses of GDP. As a key indicator of the economic performance of the market economy, it is of essential use to conduct economic policy; yet it cannot be used to measure societal progress. Indicators for noneconomic dimensions of development—such as poverty and social exclusion, risks and sustainability, and care for the environment—must complement GDP to avoid policy missteps. Alternative concepts have been tested and are increasingly used for policymaking at regional, national, and international levels.

6.2 The search for a more meaningful metric

To address the distributional shortcomings of GDP, the idea of social accounting was suggested already in the 1960s. In the following decade, several initiatives were launched aimed at correcting national accounts aggregates for environmental and other nonmarket factors. This also meant raising concern from the national level to the global arena:

- 1972: The Club of Rome published its report, *Limits to Growth*, presenting a Malthusian confrontation between limited resources and the growing population (Meadows et al., 1972).
- 1972: Participants of the UN Conference on the Human Environment held in Stockholm declared that economic development and environmental quality must be managed in a mutually beneficial way.
- 1973: Nordhaus and Tobin developed their Measure of Economic Well-being (MEW).
- 1974: Hueting proposed the Sustainable National Income (SNI) measure.

There is universal acknowledgment that GDP as a measure of production does not reflect issues such as environmental sustainability and social inclusion. It is perfectly natural within any statistical system that a metric developed for one particular purpose will be ill-suited for another. Confusion might then easily occur. The basic question is: What do analysts and policymakers want to measure? While GDP is useful to measure the levels of market activity, it is not a measure of well-being. Clearly, a set of indicators to measure social and environmental progress (distinct from economic performance) requires tools that reflect the multidimensional nature of the notion of "progress." Typically, collections of indicators to measure "progress" describe average conditions of people living in different countries and regions. The timing of measurement is also important, i.e., the given interval (such as a year) for measuring flows or changes in those flows.

The concept of "progress" made an international breakthrough with The Brundtland Report, *Our Common Future* (UNWCED, 1987), which refined the conceptualization of sustainable development and widened it beyond environmental concerns to include social aspects at national and international levels. At the same time, consideration of environmental problems led to a review of the naturally acidified environment and its effect on health. With an impressive progress in providing real time data on sustainability, the timely availability of some social and environmental indicators is still a major concern, and these efforts have not yet been incorporated into commonly accepted macroindicators.

6.2.1 The human development index and beyond

In 1990, another major milestone was realized in the search for alternatives or complements to GDP when the UN created the "human development index" (HDI). While it is a quite simple index that synthesizes only a limited amount of information, the HDI has remained one of the few regularly compiled indexes used by international organizations for systematic cross-country comparisons. To balance comparisons based on GDP per capita in purchasing power parities, HDI extends the analysis to include two major dimensions of development and progress—a nation's *health* and its *educational achievement*. Health is measured by life expectancy at birth; educational achievement is measured through the average adult literacy rate (with the weight 2/3) and the student enrollment rate (weight 1/3). The three dimensions (economic prosperity as GDP, health, and education) are then given equal weights in the overall index. Since its first publication, possible modifications to these variables have been widely discussed.

The UNDP is committed to further develop the HDI as a general index of sustainable human development, and the experience gained from constructing the HDI has played an important role in raising international awareness of essential noneconomic dimensions of the quality of life. This set the stage for the UN summit held in Rio de Janeiro in 1992 to introduce the notion of sustainable development into the policy debate ("Agenda 21"). The need for statistical documentation also became clear. Today, an updated version of the HDI includes measures of the harmonious or consistent human development of a nation beyond longevity, knowledge, and the standard of living.

Since the 1992 summit, international and supranational organizations such as the European Union, the European Commission (Eurostat), the Organization for Economic Cooperation and Development (OECD), and the World Bank have launched large-scale projects to measure sustainable development or societal progress. A list of major international initiatives undertaken to measure sustainable development and related concepts is presented below:

- 1992: Commission for Sustainable Development (UN)
- 2001: European Union Strategy for Sustainable Development (EC)
- 2005: Working Group on Statistics for Sustainable Development UNE CE-OECD-Eurostat)
- 2007: GDP and Beyond (EC)
- 2008: Measurement work on sustainable development, well-being, and social progress (OECD)
- 2009: Stiglitz-Sen-Fitoussi Report (France)

- 2010: Sponsorship group on measuring progress, well-being, and sustainable development (EC)
- 2011: BRAINPOoL
- 2012: E-frame (EC)
- 2012: Rio +20 Conference (UN)

In 2001, the OECD first addressed measurement challenges in the field of sustainable development and well-being. A report titled, *Sustainable Development—Critical Issues*, summarizes the results of a three-year project. In 2003, the OECD, with the UN, the EC, and the IMF published a *Handbook on Integrated Environmental and Economic Accounting*, proposing an accounting framework and methodologies for correcting national accounts for environmental degradation and resource depletion. In 2006, a further OECD report, *Alternative Measures of Well-being*, discussed the limits of GDP as a welfare metric. The OECD also organized a series of conferences on "Statistics, Knowledge, and Policies" (at Palermo in 2004, Busan in 2009, and New Delhi in 2012) aiming to shape a global consensus on the need to strengthen statistical work in this field. Following a 2007 forum in Istanbul, the OECD-hosted "Global Project on Measuring the Progress of Societies" was established.

In November 2007, the European Commission jointly with the European Parliament, the Club of Rome, the Worldwide Fund for Nature, and the OECD organized the "Beyond GDP Conference" with the participation of the major stakeholders in the EU institutions, its member states, the research community, statistical offices, and civil society—where a consensus was reached on the need for more adequate metrics.

6.2.2 The Sarkozy Commission

With the increasing integration and globalization of national economies as manifested by large inward foreign direct investments, capital and trade flows, and intensified labor mobility, the adjustments of GDP for the relations of a nation with the rest of the world have become an important issue. In February 2008, French president Nicholas Sarkozy asked J. Stiglitz, A. Sen, and J. P. Fitoussi to head a commission to address the prevailing dissatisfaction with the present state of statistical information about the economy and society. This "Commission on the Measurement of Economic Performance and Social Progress" or Sarkozy Commission, was charged to identify the limits of GDP as an indicator of economic performance and social progress, examine the problems in measuring GDP, and consider which additional information might deliver more relevant indicators of social progress.

According to the Sarkozy commission, the conventional measure of the value-added produced by resident units (or GDP per capita) is an imperfect proxy of the income at their disposal, indicating instead, how much economic agents can consume immediately or invest in the future. In countries employing a large community of migrant workers who transfer their wages to their respective homelands, GDP and gross national income (GNI) can differ substantially. The same applies to property income, dividends, and interest arising from domestic production that is then transferred to nonresident owners of the domestic productive units.

Calculations of GNI and gross disposable income (GDI) of an economy can also be problematic. Formally, GNI is defined as GDP minus net taxes on production and imports, and compensation of employees and property income payable to the rest of the world, plus the corresponding items receivable from the rest of the world. To adequately reflect the GDP of the economy, other current transfers to and from the rest of the world should also be taken into account, including the payments of taxes on income and property, social contributions, and social benefits.

In its report, the Sarkozy Commission presented the strengths and weaknesses of the current definition of GDP. As noted by the Commission, the inadequacy of GDP as a welfare indicator and as an index for measuring socioeconomic progress has commonly been referred to in the public debate. There is a widespread acknowledgment that GDP suffers from two major weaknesses (Afsa et al., 2008):

- Being a monetary aggregate, GDP pays little or no attention to distributional issues, elements of human activity, or well-being (for which no direct or indirect market valuation is available).
- GDP measures productive flows and, as such, ignores the impact of productive activities on stocks, including stocks of natural resources.

6.3 Recent progress

The UN has regularly reviewed a *system of national accounts* (SNA)—a consideration of acknowledged national strengths and weaknesses—and drawing on research, practical experience, and international standards for business and public accounting. The improved "SNA 2008" standard has features that fall into five groups: assets, the financial sector, globalization and related issues, the general government and public sectors, and the informal sector. This standardized approach ensures comparability between countries for global research and analysis.

In 2009, the European Commission issued the document, *A Roadmap for Action: GDP and Beyond—Measuring Progress in a Changing World,* setting out short- and medium-term actions that might lead to the forming of indicators that complement GDP and provide the knowledge base for effective and coherent policymaking:

- *Action 1*: As GDP growth figures do not provide sufficient information about social and environmental outcomes, top-level indicators on environmental protection and quality-of-life should be developed.
- *Action 2*: For policymaking and public debate, complementary environmental and social indicators need to be available together with GDP; the timeliness of such indicators needs to be improved.
- *Action 3*: GDP per capita and its growth over time does not tell how economic benefits are shared within a society. To enable a debate on equity and fairness, it is necessary to know the distribution between regions and social groups of income, health, education, and environmental quality.
- *Action 4*: GDP growth rates do not indicate whether economic development is sustainable. Tools need to be developed to measure and communicate overall sustainability.
- *Action 5*: Traditional macroindicators such as GDP do not measure environmental or social outcomes. Current accounting should be expanded to include environmental and social issues.

In 2011 in the context of the OECD's 50th anniversary, the "OECD Better Life Initiative" was launched. Its main outputs were presented in a report, *How's Life?* that included a list of well-being indicators based on a framework structured around 11 dimensions (housing, income, jobs, education, environment, civic engagement, health, life satisfaction, safety, and work-life balance). A web-based interactive tool, *Your Better Life Index*, was launched at the same time. Further work to advance the measuring of well-being and green growth is underway under the auspices of the OECD Committee on Statistics.

On the academic side, economists have shown an interest in constructing measures of well-being, life satisfaction, and happiness. Recent literature on the subject recognizes that the relation between subjective well-being and monetary income is only partial. Current international initiatives include the OECD "Better Life" initiative (measuring well-being and societal progress), its "Green Growth" strategy, and the program "Toward a Green Economy." There is a World Health Organization initiative on the "Measurement of and Target-setting for Well-being." In June 2012, the OECD, in collaboration with Eurostat and the European Statistical System, organized a European conference on "Measuring Well-being and Fostering the Progress of Societies." Some recent indicator sets cover the subject of sustainable development, even employing participatory approaches at the local level.

The fourth OECD World Forum on "Measuring Progress for Well-being and Development" in New Delhi in October 2012 similarly confirmed the desire to measure societal progress for people's sustainable well-being. Also in 2012, the UN Statistical Commission adopted the System of Environmental Economic Accounting as the international statistical standard—a multipurpose conceptual framework for understanding interactions between the economy and the environment, and for describing stocks and changes in stocks of environmental assets.

Thus the search for a more meaningful metric continues to move forward. A major outcome of the 2012 Rio +20 Conference was the launching of a process to develop a set of Sustainable Development Goals for the year 2015 and beyond, building upon the experience gained from the millennium development goals.[1] The document, *The Future We Want*, (UN, 2012b) summed up the results of the Rio +20 conference on sustainable growth and included a section *Beyond GDP*. Paragraph 38 addressed this message to the official statistical community:

> *We recognize the need for broader measures of progress to complement gross domestic product in order to better inform policy decisions, and in this regard we request the United Nations Statistical Commission, in consultation with relevant United Nations system entities and other relevant organizations, to launch a program of work in this area, building on existing initiatives.*

To meet this request, the UN Statistical Commission at its forty-fourth session (26 February to 1 March 2013) set up a "Friends of the Chair Group on Broader Measures of Progress (FOC). The Commission requested the FOC among other things to "build a work programme to develop broader measures of progress" to complement GDP in order to better inform policy decisions. The commission was tasked to work on global-level indicators to complement GDP and ensure that a robust statistical measurement approach is incorporated from the outset in preparations for the post-2015 development agenda.

[1] The World Bank on the same occasion launched an initiative to account for the natural capital resources of the world.

6.4 An indicator pyramid

Indicators have traditionally been used as tools to monitor progress in economic, social, and environmental areas and for informed policy target-setting. In the publication, Communication on Europe 2020, issued by the European Economic and Social Committee (EESC) in 2013, indicators are defined as "quantitative variables measured or calculated which, individually or as part of a set, allow obtaining as much valid empirical information as possible on a specific matter." Traditionally, the responsibility for indicator definition and use lies with the policymakers; the statistical system only guarantees the quality of the underlying data.

An indicator set consists of indicators showing the reality in all its complexity: the state of society, its social, economic, and ecological connections, and its development over time and space. When measuring an issue as broad and complex as societal progress, the EESC presents the indicator construction as a process involving various actors that necessarily cope with potentially conflicting goals of different disciplines: a balance must be struck between statistical measurability, scientific consistency, credibility/objectivity, and political relevance.

At the base of the indicator pyramid, there are the *basic data* on selected specific topics, which are collected by subject-related surveys or—as in the case of the environment—physical measurements. The basic data can be used in many ways for specialized planning, research, and monitoring purposes. At the middle level of the pyramid are *indicators* that are more aggregated or more selective in terms of their subject matter; they may be integrated into sets of indicators, e.g., subindicators. The top of the data pyramid is occupied by *macroindicators* (sometimes referred to as summary indicators). Key aggregates and composite indicators at the top reflect specific statistical information.

6.5 Composite indicators

Composite indicators can be useful for comprehensive analyses of trends of socioeconomic conditions within and between countries. A type of macroindicator, composite indicators are traditionally used to help measure multidimensional and complex phenomena—by replacing nonmeasurable phenomena with intermediate, but measurable, objectives.

A composite indicator is formed by aggregating several elementary indexes over a broad spectrum of dimensions describing the subject (whether human development, well-being, environmental sustainability, etc.) The component indicators have no common unit of measurement and no obvious weighting method usually exists. The distinctive features of composite indicators relate to the domains covered, the normalization methodology used, and the weights used for aggregation.

The 2008 OECD *Handbook on Constructing Composite Indicators* provides detailed instructions for constructing composite indicators and discusses methodological approaches for constructing aggregate measures. It has been prepared jointly by the OECD (the Statistics Directorate and the Directorate for Science, Technology and Industry) and the Applied Statistics and Econometrics Unit of the Joint Research Centre (JRC) of the European Commission in Ispra, Italy. The Handbook offers a menu of statistical methodologies and technical guidelines that can help constructors of composite indicators to improve the quality of their outputs and refers to the use of Data Envelopment Analysis for that purpose. For each step, decisions need to be made, based on expert judgment. The last three decades have seen remarkable advance in methodology and statistical work to measure sustainable development. A number of meaningful indicators have been developed by international institutions and nongovernment organizations. Composite indicators proposed in the academic literature also help place the measurement of sustainable development high on the agenda of national and international institutions. A recent inventory cataloged almost 300 composite indicators from the fields of economics, social studies, the environment, globalization, and technology. Some influential composite indicators are:

- 1989: Index of Sustainable Economic Welfare (Daly and Cobb, 1989)
- 1992: Ecological Footprint (Rees, 1992)
- 1993: Genuine Savings (Pearce and Atkinson, 1993)
- 1995: Genuine Progress Indicator (Cobb et al., 1995.)
- 1999: Sustainable Net Benefit Indicator (Lawn and Sanders, 1999)
- 2002: Index of Economic Well-being (Osberg and Sharpe, 2002)
- 2005: Environmental Sustainability Index (ESI) and an Environmental Performance Index (EPI) (Esty et al., 2005.)
- 2005: Living Planet Index (LPI) (Loh et al., 2005.)
- 2008: Sustainable Society Index (SSI) (van der Kerk and Manuel, 2008)
- 2010: Happy Planet Index (HPI) (Marks, 2010)

To cope with existing differences, the UN Commission for Europe (acting through its statistical office Eurostat) and the OECD joined efforts to establish a task force to develop a harmonized conceptual framework to measure sustainable development. On May 27, 2013, the task force issued *A Framework and Suggested Indicators to Measure Sustainable Development*— a report that was endorsed the following month by the Conference of European Statisticians (a body representing about 60 countries). The proposed framework organizes a large number of indicators and presents the information in a concise manner for statistical experts, development practitioners, and policymakers.

While composite indicators are popular among academia and environmentalist groups, policymakers' opinions diverge between the use of indicator sets *and* the use of composite indicators to guide policies. The Commission task force underscored the importance of transboundary impacts at the current stage of globalization, stressing that relationships between countries are becoming more and more important. It stated that a major challenge in identifying an appropriate approach to measuring sustainable development lies in choosing between composite indicators and indicator sets. Except for the World Bank, which relies on composite monetary indicators in its research on sustainable development (World Bank, 2011), nearly all international organizations and national statistical offices currently use indicator sets.

6.6 Empirical production/transformation functions

The use of aggregate production functions has been the major "hardware" in the theory of economic growth. The production function is an abstract mathematical relationship that describes the quantity of output as a function of the quantity of input, assuming technical efficiency—the maximum output possible for every feasible combination of inputs. It relates a nation's output of goods and services to the input of capital and labor, identifying the accumulation of physical capital, investment in human capital, and technological progress as the driving forces behind aggregate production.

Recent developments identify the important determinants of the productive process to include: management practices, geographical factors, protection of property rights, public policies toward market regulation and macroeconomic stabilization, social insurance, and conflict management. But while such institutional determinants shape the "rules of the game," they are not captured by the traditional production function. Beyond the notion of a production function, links of social factors to human well-being also need to be considered. A generalized and multidimensional "transformation function" would relate a nation's development outcomes to various inputs that go beyond capital and labor (management practice, institutions, etc.) Yet the exact functional relationship between such inputs and the outputs may be unknown, even while clear empirical evidence supports such transformation.

In conventional microeconomic theory, the specification of a production function determines the nature of the input-output relationship in an organization. The underlying assumption is then the existence of a functional transformation technology that determines what maximum amount of outputs can be produced from a given combination of inputs (or alternatively, the minimum combination of various inputs needed to produce a given level of outputs). However, as observed by Seiford and Thrall (1990), "This description of the production technology would be provided by the production function, if it were known." In reality, the production/transformation function is not known.

A distinguishing feature of data envelopment analysis (DEA) is an absence of assumptions about the underlying functional form relating independent and dependent variables. No theoretical relationship is assumed for DEA. It is a "black-box" technique that can be used when there is insufficient knowledge or lack of consensus on the nature of causal relationships. Only data are needed—observations about the various inputs and their magnitudes and about the resulting output variables and their magnitudes. DEA constructs a piecewise empirical production frontier from the observed data. DEA can handle multiple inputs and outputs that are interacting in complex and unknown ways and may be measured in different units.

By "empirical" function (Charnes et al., 1983), we shall mean a vector function whose values are known at a definite number of points and whose values at other points in its domain are constructed by linear (usually convex) combinations of values at known points. The points in the domain are "inputs," the component values of the vector function "outputs."

6.7 Why data envelopment analysis?

DEA sharpens the evaluative features of mathematical programming to assess actual accomplishments and deliver superior management information. DEA is an analytical system acclaimed to be "a leading-edge method of performance measurement that supports benchmarking, continuous improvement, and strategic analysis." One can only agree with Bouysou (1999) that "DEA can safely be considered as one of the recent success stories" in operations research.

For the purpose of DEA, observations are collected for a finite number of entities called decision-making units (DMUs), such as private firms, hospitals, or policy-making units (PMUs) (as in our case) entire countries, cities, regions, etc. The construction of a typical DEA model will then involve the following steps:

- determining the input and output variables;
- choice of optimization orientation—input minimization or output maximization (in the present volume output maximization is usually assumed);
- possible weight restriction (see further below);
- the use of cross-sectional data or longitudinal data.

The list below enumerates reasons why DEA is well-suited to be used for policy impact assessment at the national or even global level:

- DEA reveals frontier rather than central tendencies.
- DEA does not require explicit assumptions about functional forms or relative weights to obtain its results.
- DEA copes with information asymmetries; it is useful for uncovering relationships that would remain hidden from other methodologies.
- DEA can handle multiple noncommensurate socioeconomic variables interacting in complex and unknown ways, when there is insufficient knowledge of causal relationships or a lack of consensus on alternatives. DEA can be used to address managerial and economic issues providing useful results for conducting *ex-post* evaluations of the performance of the DMUs/PMUs, and thus enable decision makers to agree upon *ex-ante* measures directed at improving their performance.
- DEA is an integrated approach to multiple criteria decision making. It establishes benchmarks, measures DMU/PMU performance compared to these benchmarks, and informs how benchmarks can best be reached.
- DEA will provide answers to the following questions:
  - Who are the best-in-class decision makers managing to perform better under the same conditions?
  - How well is a decision maker doing relative to the others—not relative to a theoretical maximum but in *real life*?
  - Whom could a decision maker learn from?
  - What would be its target levels of observed parameter values?
  - Where and how much could it improve?

DEA is a benchmarking tool that evaluates a population of DMUs/PMUs (countries in our case) in their performance in converting input to outputs. The purpose of the analysis is to identify the countries that most effectively transform their inputs to outputs. These units are located on the effectiveness frontier. The remaining units, falling behind the frontier, are deemed ineffective.

The efficiency score of DEA is 1.00 for frontier points; the virtual output then equals virtual input. It is less than one for subefficient points located behind the frontier (virtual output falling short of virtual input). The frontier (envelope) itself consists of the efficient observations (with scores equal to one) and the surface of the convex hull enveloping them. Frontier points are referred to as "best practice" or "benchmarks." Any best practice point can be written as a linear combination of its "reference" units or "peers" (e.g., the corner points of the current frontier facet). For a country that falls behind the frontier, DEA calculates the additional output that the unit might have produced with best practice, which is the target for an inefficient unit—found by projecting the inefficient unit radially to the frontier (stepping up all outputs in the same proportion while using the same inputs). The peers identify "role models" from which a country should be able to learn how to improve its performance. DEA assumes that data are error free and that performance differences of the DMUs/PMUs are due to different resource mixes and managerial effectiveness.

In conclusion, DEA's efficiency scores and the envelopment frontier are precise—based on real observation, rather than estimations. DEA calculations focus on individual observations in contrast to population averages, and they produce a single aggregate measure for each country in terms of its utilization of input factors (independent variables) to produce desired outputs (dependent variables). Other attractive features include:

- DEA can simultaneously utilize multiple outputs and multiple inputs, each being scaled in its own units;
- DEA can be adjusted for the presence of exogenous variables, and it can incorporate categorical variables;
- DEA calculations are value-free and do not require specification or knowledge of any priory weights or prices for the inputs or outputs;
- DEA makes no assumptions regarding the functional form of production/transformation relationship;
- DEA can accommodate personal judgments when so desired.

CHAPTER SEVEN

Beyond the Washington Consensus

We need a new Geneva consensus ... for the opening up of trade that takes into account the resultant cost of adjustment ....We need to reform globalization with a clear view to enhancing the development of social, economic and ecological aspects of humanity.... Trade opening is necessary, but it is not sufficient in itself.... Humanizing globalization means that we must take care of the victims of globalization. To achieve this, global solutions must be sought for addressing the negative impact of globalization at all levels—individual, community, and universal levels.

(Pascal Lamy, *Director-General, WTO, Santiago 2006*)

7.1 The Washington Consensus

In world history, the 1980s will remain the decade of three world leaders: Thatcher, Reagan, and Gorbachev. Together they demonstrated the importance of personal leadership for the future of nations and for humanity at large. The liberalization and deregulation policies they initiated and pursued launched a new course in international affairs: the era of globalization. With the collapse of the Soviet empire and the disintegration of the former Council for Mutual Economic Assistance that had been founded in 1949 as an alternative to the Marshall plan for the Soviet satellite states in Eastern Europe, the bankruptcy of the Soviet state-regulated socialist economic system became apparent to everybody. Under the aegis of the United Nations (UN), the newly independent countries joined the large family of international institutions, embarking on a process of accession to the World Trade Organization's General Agreement on Tariffs and Trade (GATT) to be able to benefit from the fruits of open global markets. Inspired by the triumph of market-oriented economic theories over theories advocating government ownership or control, a widespread process of privatization, deregulation, liberalization, competition, and agreements to lower trade barriers commenced globally.

The relationship between globalization and development has been one of the most challenged issues in economic policy in recent decades. The Washington Consensus as formulated by the British economist John

Measuring Sustainable Development Goals Performance
ISBN: 978-0-323-90268-7
https://doi.org/10.1016/B978-0-323-90268-7.00006-2

Williamson (1990) refers to a set of broadly free market economic ideas such as free trade, floating exchange rates, free markets, and macroeconomic stability. At the time, the consensus expressed the general economic philosophy of prominent economists and international organizations in Washington DC including the IMF, the World Bank, and the US Treasury. The underlying assumption of the consensus was that trade openness is beneficial and indeed essential for the growth and development of a developing country. In order to open such economies to favorable development, it was suggested that they liberalize their imports and orient production toward exports to ensure faster growth.

The term was coined in 1989 in a background paper presented by Williamson for a conference that the Institute for International Economics convened to examine the extent to which old prescriptions of development economics that had governed Latin American economic policy since the 1950s, were being swept aside by a new set of ideas that had long been deemed appropriate within the OECD. Williamson presented a list of 10 groups of policy interventions (10 commandments) that he believed practically everyone in Washington agreed were needed to be implemented in Latin America. This, then, was the Washington Consensus. By implication, senior members of the Congress, the administration, the international financial institutions in Washington, the economic agencies of the US government, the Federal Reserve Board, and the think tanks would all subscribe to these ideas.

The policy reforms aimed at initiating industrialization and import substitution. According to Williamson, the original conception of the consensus was founded on three big ideas: a market economy, trade liberalization and openness to the world, and macroeconomic discipline. The 10 reform agenda items are:

(1) fiscal discipline to eliminate public deficits;
(2) a change in the priorities of public spending: withdrawal of subsidies and increased spending on health and education;
(3) tax reform: broadening tax bases and reducing tax rates;
(4) positive real interest rates determined by the market;
(5) exchange rates determined by the market guaranteeing their effectiveness;
(6) liberalization of trade and opening of the economy (remarkably, however, Williamson did not include the liberalization of capital flows);
(7) no restrictions on foreign direct investment;
(8) privatization of public enterprises;

(9) deregulation of economic activity;
(10) a solid guarantee of property rights.

Elaborating on these points, Williamson later (2008) made the following additional observations:

- *Fiscal discipline*: At the time, almost all countries in Latin America had run large government deficits leading to balance of payments crises and high inflation. As a result, considerable capital flight to abroad was occurring, deepening domestic levels of poverty.
- *Reordering public expenditure priorities*: This suggested switching public spending patterns toward policy measures that would both stimulate growth and alleviate poverty, from nonmerited subsidies to basic health care, education, and infrastructure.
- *Tax reform*: The aim would be a tax system combining a broad tax base with moderate marginal tax rates.
- *Liberalizing interest rates*: Recognizing the importance of financial liberalization but saying very little about the necessity for prudential supervision of banks and financial institutions.
- *Competitive exchange rates*: Asserting that exchange rates should be left free and without interventions from the central bank.
- *Trade liberalization*: This was clearly one of the key principles of the policy recommendations. Only scant attention was given to the issue of how quickly borders should be opened up and trade be liberalized.
- *Liberalization of inward foreign direct investment*: Under this heading Williamson never included comprehensive capital account liberalization, believing that it did not or should not command a consensus in Washington.
- *Privatization*: An originally neoliberal idea that at this time seemed to have won broad acceptance within the Washington community.
- *Deregulation*: This focused specifically on easing barriers to entry and exit, not on abolishing regulations designed for safety or environmental, nor to govern prices in a noncompetitive industry.
- *Property rights*: This was primarily about providing the informal sector with the ability to gain property rights at acceptable cost.

The 10 categories of recommendations of the Washington Consensus can be organized into two mutually reinforcing groups of policies: stability through fiscal adjustment and market orthodoxy (commandments i, ii, iii, x); and liberalization through a dramatic reduction of the role of the state in the economy (iv, v vi, vii, viii, ix).

7.2 The Washington Consensus criticized

Since its inception, the term Washington Consensus has provoked ideological controversies. In 2008, Narcís Serra and Joseph Stiglitz edited a volume titled, *The Washington Consensus Reconsidered: Toward a New Global Governance*. The book discusses the relationship between the articles of the consensus and how the term has come to be widely understood. The team of authors provides a comprehensive discussion of the shortcomings of the suggested reforms. They conclude that both theory and evidence weigh heavily against what has come to be called the Washington Consensus.

The original intention of the Consensus was to formalize a set of development strategies that commanded a consensus in some significant parts of Washington. It was inspired by the liberalization policy instruments of the Bretton Woods institutions and the World Trade Organization. For many critics, however, it seemed to discount the role of government policy and move toward a global *laissez-faire* economy where the only thing that mattered would be the growth of gross domestic product (GDP).

An early international development strategy of the 1970s expressed concerns with the environment and called for intensified national and international efforts, searching "to make the world, in its economic, and social aspects, a more just and rational home for mankind." As a result, in 1973, the UN initiated a system-wide project to study environmental aspects of the future world economy. In the UN General Assembly resolution 3201 (S-VI) of May 1, 1974, an alarming and growing gap between developed and developing countries was established, calling for corrections of inequalities and redressing existing injustices, with the aim of narrowing the gap and ensuring an acceleration of economic and social development and justice for present and future generation.

In 1977, the UN published a study titled, *The Future of the World Economy*, presenting the impact of prospective economic issues and policies on the International Development Strategy for the second UN Development Decade (Leontif et al., 1977). The Washington Consensus, against this background, many considered it to have disregarded both the advantages of industrial policy and reasonable equity considerations (supporting an adequate legal and institutional safety net).

The Washington Consensus was predicated upon a strong faith in unfettered markets and aimed at reducing, or even minimizing, the role of government. To its critics, the Consensus has come to be associated with market fundamentalism—the view that markets solve most, if not all, economic problems by themselves. Already in the 1970s, however, advances in economic theory had shown that markets can fail due to imperfections in information, limitations in competition, and incompleteness. Markets do not always produce efficient outcomes by themselves, implying the necessity of a role for government intervention. According to Stiglitz, these advances in economic theory had already removed the intellectual foundations of market fundamentalism long before the Washington Consensus became fashionable.

The dividends of macroeconomic stability are still open to controversial interpretations. A broader view of macroeconomic stability would include not only price stability and sound fiscal policies but also the sustainability of the real economy. To its detractors, neither theory nor practice supported the plausibility of the Washington Consensus. Accordingly, it should not have come as much of a surprise that the Consensus prescriptions failed to work as promised. Disillusion with the Washington Consensus grew throughout the developing world. The agenda seemed to many neither necessary nor sufficient preconditions for sustainable growth, ultimately failing to deliver policies that would promote the advancement of developing countries.

In retrospect, Williamson concluded that there was no consensus around the Washington Consensus after all, but rather a convergence of opinion around key concepts. He suggested that it would have been more appropriate to label his list the Universal Convergence rather than a consensus, "since the extent of agreement is far short of consensus but runs far wider than Washington."

7.3 Post-Washington Consensus

In the aftermath of the Washington Consensus, there is a general agreement about the nature of many of its shortcomings. The most important of these is a failure to consider local economic structures within developing countries, focusing on a too narrow set of objectives and on a too limited set of policy instruments. The development goals determining the policy framework should not be just an increase in GDP, but should regard the distributional impact of policies that would look to include equitable and sustainable increases in societal well-being.

The Washington Consensus clearly failed to adequately address the subject of governance—the role and character of public institutions that has later been placed at center stage, stressing not only the importance of governance of individual countries but also the issue of global governance. To many, the Consensus had confused ends with means, so that privatization and liberalization had become not means to an end, but ends in themselves. Major development goals such as reducing poverty, building equity, and sustaining the environment were not part of the Consensus protocol. Stiglitz identifies six elements of a post-Washington Consensus:

(1) a post-Washington Consensus cannot be arrived at simply within the confines of Washington;
(2) one-size-fits-all policies are doomed to fail;
(3) economic science has not yet provided sufficient evidence, sufficiently strong theory, or empirical evidence, to result in a broad consensus about what countries should do;
(4) a critical complementary role of government with markets associated with the developmental state;
(5) development requires strengthening of both market and state institutions;
(6) success is to be measured not just by an increase in GDP, but by a broader set of measures—including those that assess environmental and social sustainability.

7.4 The Barcelona Development Agenda

In response to widespread criticism of the Washington Consensus, the Universal Forum of Cultures met in Barcelona in September 2004 to consider the prospects for growth and development around the world (The list of participants included both Williamson and Stiglitz.). The Forum proposed an alternative system of policies featuring seven general principles that were believed to offer more flexible approaches to development.

The Barcelona Development Agenda (UFC, 2004) spelled out a broader concern for equitable and sustainable development as reflected in long-term goals for macroeconomic policy, a wider range of economic policy instruments, a balanced role for markets and government, and the creation of a better system of global governance. The conference proposed reforms both at national levels and at the global level responding to two sets of questions: What can each country, individually, do to enhance sustainable, stable, equitable, and democratic development? and How should the global economic architecture be changed? There was agreement among the participants on seven "sets of lessons" that would serve as priorities for reform:

(1) *Institutional quality and market orientation matter.* The rule of law and property rights is a decisive precondition for successful development to occur. A country's history, culture, and other specific circumstances need to be respected.

(2) *Poor financial policies are barriers to development.* Developing nations that hope to prosper should pursue prudent financial, monetary, fiscal, and debt policies—as weak and maladjusted measures manifest in large debts, both public and private, while poorly regulated banks and loose monetary policies are serious threats and tend to inhibit development. Large public debt can burden governments with debt service obligations that consume a significant proportion of the GDP, and might result in financial crises that disproportionally affect low-income households.

(3) *Sustained growth cannot be guaranteed by a single set of policies.* To ignite sustained growth, international lending organizations and aid agencies should encourage countries' ownership over their development agenda and shall be free to experiment with policies that would suit their specific circumstances best. The most binding constraints to growth shall be identified and addressed through a combination of microeconomic, industrial, and macroeconomic policy mixes.

(4) *Multilateral trade negotiations should proceed in a manner that promotes development.* Since agricultural and textile protectionism in developed countries creates obstacles for market entry and expansion in developing countries, trade negotiations should include stakeholders from impacted states. Trade policies in developing countries should be examined to identify and correct those that limit growth. A successful conclusion of the Doha Round[1] will provide more opportunities for world growth, thereby creating more room for developing countries to pursue their own growth strategies.

(5) *Current international financial relationships are not working well.* The international financial arrangements should be reformed. Both private capital and development financial flows to middle-income countries are highly volatile and delinked from economic fundamentals in the recipient countries. Developing nations are underrepresented in the

[1] The Doha Round is the latest round of trade negotiations among the WTO membership. Its aim is to achieve major reform of the international trading system through the introduction of lower trade barriers and revised trade rules.

decision-making bodies of the multilateral lending institutions. Their voting power is outdated, reflecting the relations of the past.

(6) *Asymmetric international arrangements on movements of capital and labor.* Current policies of international financial institutions and G7 governments generally favor the transfer of financial capital, but often discourage the movement of labor capital across borders. Both equity and efficiency considerations support allowing greater international migration. A set of international rules and institutions are needed to guide cross-border movements of people, improving the rights of migrants.

(7) *The worsening of the environment, including problems of global warming, need to be tackled.* One billion human beings living in poverty should be a cause for unrelenting concern. AIDS and other epidemic diseases represent a tragedy for the least developed countries. Donor nations' commitment to increase aid levels to meet the Millennium Development Goals remains largely unfulfilled.

There was also a recognition that the international community had to address effectively other important economic problems such as exchange rates instability, problems with the global reserve system, and the possibility of sovereign defaults.

The Barcelona agenda represents a major departure from the past, stressing the importance of a fair income distribution, poverty reduction, protection of the environment, as well as tailoring policies to country-specific situations. It recognizes that the complex relationship between globalization and sustainable development calls for an integrated multi-disciplinary analysis of the policy instruments with respect to timing, the consequences of reforms on the economic security of developing countries, and with due consideration of national sovereignty.

7.5 The way forward

The declaration of the *Rights of Man and of the Citizen* dating from the French revolution in 1789 has served as a founding stone in the history of universal human rights. It is valid at all times and in every place and it pertains to human nature itself. Transferring the concept of liberty to the sphere of economics, the idea of "economic liberalization" can only be ethically and politically defended provided that such liberty does not injure others (or at least minimizes possible negative impacts on others), that it applies to all members of society (or, failing this, that timely compensatory mechanisms are established), and that the limits of economic liberalization are defined by law or existing social and physical infrastructure. As a consequence, even a liberal economic regime needs institutional support and state regulations.

The Washington Consensus can best be characterized as a consensus for liberalization and globalization. It was not a consensus for equitable growth and sustainable development. Unfortunately, economic globalization has outpaced political globalization and global governance has not kept pace with the needs of the current version of a globalized world with universal equal rights for all. This calls for reform of multilateral arrangements toward a more coherent and just system.

Many feel that the international rules of the game are often unfair, and the international institutions have pushed one particular ideology—an ideology that has resulted in economic policies particularly ill-suited to many developing countries. Stiglitz points his finger at an undemocratic governance of the international institutions: existing voting structures do not reflect current economic standing of countries. Given such asymmetry in the governance structures, it is unsurprising that the decisions and actions made by international economic institutions reflect the ideology of advanced industrial countries, and even the interests of their multinational corporations and financial institutions.

In order to be effective, the International Monetary Fund must be seen as representing the interests of all its 188 member countries, and that its governance structure reflect the realities of today's world economy. In November 2010, the IMF agreed to wide-ranging governance reforms reflecting the increasing importance of emerging market countries. Currently IMF member states are in the process of ratifying this 2010 Quota and Governance Reform. The reforms will ensure that smaller developing countries can retain influence on the operations of IMF—a major achievement in enhancing the credibility, legitimacy, and effectiveness of the Fund. All EU member states have already fully ratified the Quota and Governance Reform.

In January 2013, the Executive Board of the IMF further formulated important building blocks for a revised quota formula and agreed on a review of the formula together with the 15th General Review of Quotas by January 2014. The review is expected to reflect the changing dynamics in the global world and will lead to further increases in the quota shares of emerging market economies, producing a combined shift of 9% of quota shares to these countries.

The World Bank governance reform process is less advanced. Some progress has been made on the subject of voting reform but more is needed in terms of finding an appropriate voting formula by 2015. With the aim of ending extreme poverty and promoting shared prosperity and environmental sustainability, the World Bank put forward a Development Committee paper titled, *A Common Vision for the World Bank Group*, outlining the proposed goals and principles for the future strategy of the World Bank Group (World Bank and IMF, 2013).

The *Oxford Handbook on Governance [Levi-Faur]* published in 2012, outlining formalized models of governance, highlights the importance of information gathering and information processing in governance. It acknowledges that information will often be politicized prior to being acted upon. Someone must consider the governance environment, react to information collected, make decisions, and then continue to consider the consequences of each round of governing.

The Rio +20 Conference, held in June 2012, in Rio de Janeiro, was an important milestone in initiating a process of strengthening the institutional framework for policies aiming at sustainable development goals. The document, *The Future We Want* (UN, 2012b), stressed the growing recognition that we live on a single planet within explicit planetary boundaries, and in an increasingly interdependent global multidimensional system that goes well beyond the economic marketplace.

The framework of UN activities and its system-wide agenda for sustainable development is now integrated both with the work of the General Assembly and the deliberations of the Economic and Social Council (ECOSOC). This coordination has led to the replacement of the former Commission on Sustainable Development by a new High Level Political Forum (HLPF) on Sustainable Development representing a group of prominent political leaders devoted to the common cause of sustainable development. The HLPF is mandated to oversee the implementation of sustainable development commitments, provide political leadership and guidance, enhance integration of the dimensions of sustainable development within a single framework, improve cooperation and coordination within the UN system, and strengthen the science policy interface. The Forum meets annually under the auspices of the Economic and Social Council for eight days, including a three-day ministerial segment and every four years at the level of Heads of State and Government under the auspices of the General Assembly for two days.

HLPF over the years

2021—Sustainable and resilient recovery from the COVID-19 pandemic that promotes the economic, social, and environmental dimensions of sustainable development: building an inclusive and effective path for the achievement of the 2030 Agenda in the context of the decade of action and delivery for sustainable development

(SDG1, SDG2, SDG 3, SDG 8, SDG 10, SDG 12, SDG 13, SDG 16, SDG 17

2020—Accelerated action and transformative pathways: realizing the decade of action and delivery for sustainable development

2019—HLPF under the auspices of the Economic and Social Council

(SDG 4, SDG 8, SDG 10, SDG 13, SDG 16, SDG 17)

Empowering people and ensuring inclusiveness and equality

HLPF under the auspices of the General Assembly (SDG Summit)

2018—Transformation towards sustainable and resilient societies

(SDG 6, SDG 7, SDG 11, SDG 12, SDG 15, SDG 17)

2017—Eradicating poverty and promoting prosperity in a changing world

(SDG 1, SDG 2, SDG 3, SDG 5, SDG 9, SDG 14, SDG 17)

2016—Ensuring that no one is left behind

2015—Strengthening integration, implementation and review—the HLPF after 2015

2014—Achieving the MDGs and charting the way for an ambitious post-2015 development agenda, including the SDGs

2013—Building the future we want: from Rio+20 to the post-2015 development agenda

There is a hope that working at a higher political level with regular meetings, the HLPF will contribute to enhancing coherence.

CHAPTER EIGHT

Toward a sustainable globalization

I support globalization and believe it can bring more benefits to the poor than its alternative. But it must be the right kind of globalization. To me, globalization is like a hundred-lane highway criss-crossing the world. If it is a free-for-all highway, its lanes will be taken over by the giant trucks from powerful economies. Bangladeshi rickshaw will be thrown off the highway. The rule of strongest takes it all must be replaced by rules that ensure that the poorest have a place and piece of the action, without being elbowed out by the strong.

(Muhammad Yunus, *Nobel Memorial Lecture, 2006*)

8.1 Challenges of globalization

The term globalization is used in many ways, but the underlying principal is the progressive integration of economies and societies within a common global framework. It is driven by the dissemination of new technologies, expansion of new economic relationships, and the interconnectedness of national and international policies engaging an ever-growing range of international actors including governments, business, labor, civil society, and scientific and religious organizations.

A rich literature defines the phenomenon of globalization—discussing the worldwide political, socioeconomic, environmental, and cultural interdependency and awareness—that has become a major driving force for change in today's world. This change is expressed in increasingly free cross-border flow of ideas, people, goods and services, and capital—as well as environmentally and biologically harmful substances.

Broadly speaking, there are *two* major aspects of globalization. *The first* refers to factors and institutions—such as trade, technology, multinational enterprises, and communications that bring societies and citizens closer together to shape a single global marketplace. Deregulation of markets, technological innovation, and cheaper communication have allowed investors to diversify their participation in competitive overseas

Measuring Sustainable Development Goals Performance
ISBN: 978-0-323-90268-7
https://doi.org/10.1016/B978-0-323-90268-7.00009-8

markets. In consequence, a significant increase in cross-border financial movements, including direct foreign investment, has become a key factor in international economic integration.

The second refers to policies—such as trade and capital market liberalization, agreements on international property rights, technologies and communications that are pursued both at the national and the international level—enhancing the openness of and integration between countries. These policies hopefully inject fairness and accountability into international trading and financial systems aiming at a level playing field for all actors. From this perspective, the existing pattern of globalization is not an inevitable trend. It is, at least in part, the product of policy choices and, sometimes, misuses of discretionary power in the interest of the few. While technical change is usually irreversible, government policy can be changed to accommodate societal needs. Actually, technological advance can by itself widen the policy choices available for the pursuit of sustainable development goals.

A global framework is needed to insure against political risks, covering both imports and exports. Unfortunately, the governance of globalization lags far behind the accomplishments of "hard" sciences, such as the advances in information and communication technology. Against the background of the unprecedented success of the Internet currently serving about 2.5 billion persons, there is still a long way to go before the potential benefits of the Globalization WAN (wide area network) initiative will reach individual countries and their populations.

There are widely differing perceptions of the impact of globalization, especially concerning its possible favorable or destructive economic, social, and environmental consequences. This partly stems from its varying effects on different social groups and actors across the world. The wealth of individual countries and the well-being of their peoples increasingly depends on what happens outside their boundaries. Access to foreign markets, foreigners' access to their markets, and instability in exchange rates and capital markets create both opportunities and threats for sustainable development. People as global citizens form their perceptions of globalization partly based on the statistical record and also as part of the political discussion. Modern information and communication technologies offer unprecedented opportunity for the global citizenship to express its voice in shaping policy choices. Policies can be fine-tuned, adjusted, and if necessary changed, by responsive policymakers to ensure their social legitimacy.

8.2 The social dimension of globalization

Social problems such as poverty and inequality have their genesis in a preglobalization era. Some argue that globalization has exacerbated these problems while others contend that globalization has contributed to a reduction in their incidence. In either case, these social problems clearly need to be addressed if globalization is to be politically viable, economically sustainable, and ethically acceptable.

To reduce inequality between and within countries and to ensure that the benefits of the global economy benefits reach working people and their families, the International Labor Organization (ILO) in November 2001 took the initiative of establishing the World Commission on the Social Dimension of Globalization. The Commission had 25 members and was co-chaired by President Tarja Halonen of Finland and President Benjamin Mkapa of Tanzania. The Commission was established to respond to the needs of people as they cope with the unprecedented changes that globalization has brought. The social dimension of globalization, as originally defined by the Commission, would embrace the impact of globalization on the life and work of people, on their families, and their societies. Concerns and issues are often raised about the impact of globalization on employment, working conditions, income, and social protection. Beyond the world of work, the social dimension encompasses security, culture and identity, inclusion or exclusion, and the cohesiveness of families and communities.

Recent decades have been remarkable by the measures taken by governments at the national level—and the serious attempts by international regulatory bodies, funds, and programs—and a growing number of multilateral initiatives to manage the social consequences of globalization. A number of international agreements and global treaties stress global commitment to eradicate poverty, address unemployment, and promote social integration. These efforts culminated in the recent landmark report, *A New Global Partnership: Eradicate Poverty and Transform Economies Through Sustainable Development* (UN, 2013), presented in May 2013 to the UN secretary-general by the High-level Panel on the Post-2015 Development Agenda. The panel is a 27-member group of high-ranking persons established in 2012 to provide recommendations on advancing the development framework beyond the target date for the UN Millennium Development Goals. Most of these challenges still need to be addressed and implementation mechanisms still need to be identified. There have been positive developments in reducing extreme poverty, decreasing mortality rates, progress in primary school enrollment, and achieving gender equality. Contrary to the hopes and expectations of many, however, markets have failed to deliver a steadily increasing number of productive jobs throughout the world.

8.3 Widening inequalities

The past decade has revolutionized the spread of information and communication technologies, bridging differences in time and space. Widespread use of e-banking, e-commerce, and knowledge networks has decreased transaction costs and facilitated the process of acquisition and dissemination of knowledge. But it has also increased global inequalities, as the use of advanced information and communication technologies has enabled a small number of leading companies and individuals to capture a larger share of the market and thus to increase the gap between the market leaders and the followers. While modern IT has become an instrument for advancement for some economic actors, for others it has created a new kind of insecurity. Globalization reveals a growing global inequality of technological skills and know-how.

While the World Trade Organization (WTO) operates on the basis of a single vote for each country, key decisions are made by leading economic powers in closed meetings. The executive directors representing France, Germany, Japan, the Russian Federation, Saudi Arabia, the United Kingdom, and the United States account for 46% of the voting rights in the World Bank and 48% in the International Monetary Fund. As a result, the United States and the European Union are able to protect their agricultural industries to the detriment of many developing nations (UN, 2002a).

Unfortunately, the current model of globalization is widening inequalities between and within countries. The benefits of globalization are not reaching enough people. Meeting the target of sustainable development calls for a radical change of thinking. It means starting new companies, creating decent jobs, and investing in health, education, and food safety. The goal is more satisfying levels of human development for all, including the most vulnerable, and accelerating the pace of economic transformation, growth, and sustainable development.

Intensive growth of employment is needed to meet the common goal of the international community to reduce global poverty. Governments, the private sector, international financial institutions and development agencies, and civil society representatives all seem to agree that current inequality trends must be reversed and global governance structures reviewed. It is no longer feasible to conduct "business as usual" in international economic policy.

Several Nobel laureates have recognized the importance of good governance and the need for timely reform. Incomplete information, inadequate institutions, and local capacity limitations can distort the income distribution and deepen inequality. This calls for a new role for governments to complement market activities that ensure poverty reduction and improvements in the living standards of the most needy. There is a need to reorient current austerity policies from just reducing government debt, toward social and environmental accountability. The management of economic policy in individual countries needs to include the identification and consideration of "soft" issues like local values and cultural heritage. Yet such processes of cultural change call for innovative approaches, which makes the challenge of coping with globalization even more complex.

Thus the time for consensus-building and new thinking toward sustainable, gender-sensitive, people-centered development has arrived. Existing social and economic divides leverage major threats against global prosperity, security, and stability. Widely spread terrorist attacks have demonstrated that globalization is not only about economic integration. It confirmed ILO's credo that "universal and lasting peace can be established only if it is based upon social justice" and that "poverty anywhere" accompanied by inequality and marginalization "constitute a danger to prosperity everywhere" and are among the causes of today's ills: terrorism, crime, and corruption.

Such inequalities compel enhanced collaboration among global development partners and stakeholders. There is a need to identify pathways to a sustainable future through an inclusive globalization of the world economy. We should steer the liberalization of markets toward putting a "human face" on the globalization process, thus introducing a social conscience into the global economy so that markets work for all.

Despite general awareness of these mounting problems, the relationship between globalization and labor remains largely unexplored. Analytical research is thus thrust to center stage to help assess the social impact of globalization and trade liberalization.

A number of studies have analyzed the effects of globalization on the well-being of people, and the benefits of trade liberalization for the poor.

The findings of several of these studies have been challenged and on both methodological and empirical grounds. The Development Forum on Integrating Economic and Social Policies to Achieve the United Nations Development Agenda held at the UN in New York in march 2005 by the UN. Department of Economic and Social Affairs concluded that "the proposition that trade liberalization by itself leads to higher growth remains unproven" and that "trade liberalization has resulted in both an increase and a decline in the growth rate depending on country circumstances." (Ocampo et al., 2007). The single way of legitimizing globalization worldwide is to ensure a sustainable livelihood through full employment.

Some researchers have pointed out the need to improve risk management capacity at both global and national levels, including development of an "early warning system" where problems might be identified and communicated early, so that appropriate mitigating measures can be activated. Protecting the most vulnerable population groups minimizes social costs. Measures to alleviate the social costs of globalization would include improving the functioning of labor and credit markets, a social safety net, and active labor market policies to assist displaced workers.

Lessons of the past decades call for careful considerations in policy formulation. Policies should serve the needs of the majority of people—but not sacrifice the destinies of the remaining minority by experimenting with new, unproven policies. Wise development policy implements positive measures that have worked in other countries. The Political Declaration of the World Summit of Sustainable Development (UN, 2002b) in Johannesburg recognized that:

- Sustainable development requires a long-term perspective and broad-based participation in policy formulation, decision making, and implementation at all levels and a stable partnerships with the private sector, labor, civil society, and all major interest groups.
- Successful private corporations have a responsibility to contribute to the evolution of equitable, sustainable communities; the private sector should operate within a transparent, stable regulatory environment.
- Dialogue and cooperation among all the world's peoples and civilizations is of prime importance.

The Summit declared that the achievement of the goals of sustainable development entails a democratic system of global governance with enhanced and accountable international and multilateral institutions.

8.4 The Oxford Martin Commission

The report, *Now for the Long Term*, published by the Oxford Martin Commission for Future Generations (OMCFG, 2013), calls for urgent action and a radical shakeup in politics and business to deliver progress on climate change, poverty, governance, and disease. Composed of prominent global leaders and chaired by Pascal Lamy, former WTO director-general, the Commission argues that the ability to address today's global challenges is undermined by the absence of a collective societal vision and urges leaders to promote shared global values to protect the prospects for future generations. Searching for new ways to manage our collective interests, the Oxford Martin Commission claims that failure to address the long-term issues exposes current generations to unacceptable instability and risk, and threatens our ability to build a sustainable, inclusive, and resilient future for all. Based on past successes and failures, it proposes three actions to overcome deep political and cultural divides in the world:

- *Create a coalition to counteract climate change* to accelerate action with measurable targets for energy-efficient buildings, faster market penetration of efficient vehicles, and tracking harmful emissions.
- *Establish a voluntary taxation and regulatory exchange* to address tax abuse/avoidance, harmonize company taxation arrangements, promote information sharing, and enhance transparency and governance.
- *Introduce* "CyberEx," *a new early warning platform* to promote better understanding of common cyber threats, identify preventive measures, and minimize future attacks for the shared benefit of government, corporate, and individual interests.

The report asserts that achieving sustainable globalization is an imperative rather than an option and that sustainable development can only be reached through the active participation of all stakeholders—not only government agencies but also legislators, businesses, trade unions, academia, and religious and nongovernmental organizations. It points to the great challenge of combining the centralized management of governments with the decentralized decision making of market participants.

8.5 Globalization and poverty

The impact of globalization on sustainable development and poverty reduction has been the subject of extensive debate over the past decades. Despite remarkable progress in the fight against poverty in some parts of the world, efforts often ignore the underlying causes of poverty and inequality, considering them only byproducts of growth. The relationship between globalization and growth is theoretically ambiguous. Even the simpler question of how trade affects economic growth is still under debate, perhaps because growth itself is less well understood than the economics of static markets (Deardorff and Stern, 2001). While trade affects the income opportunities and levels of the poor (both positively and negatively), the linkages between trade and poverty are not as direct and immediate as with other national policies. Trade reforms will create some losers who suffer temporary adverse shocks (some even in the long run), and exacerbate poverty with spells of unemployment. Yet, appropriate policy responses can help alleviate hardship and facilitate adjustment, rather than abandoning reform (Ben-David et al., 1999).

Studies of poverty and relative deprivation are usually founded on a comparison of income against basic material needs for food, shelter, and clothing. From an early point, they have also encompassed a wider appreciation of many cumulative effects on the life of poor people, perhaps best summed up as the loss of dignity or the humiliation that poverty engenders. Dignity is, of course, much less susceptible to quantification as it requires a sense of self-respect and respect from others, as well as material well-being. Yet surveys indicate that the most injurious and debilitating characteristic of poverty is loss of dignity because it weakens relationships within families and communities and persons in authority, and thus reduces the capacity to escape from a cycle of deprivation that can endure for generations (ILO, 2001a).

The efforts of the international community in the fight against poverty have been remarkable, culminating in the Rio +20 conference. A consensus was reached to pursue poverty reduction and sustainable development through the means of employment-intensive growth. Further results reached in 2013 are significant, with a global accord signed during the WTO ministerial meeting at Doha, with the Monterrey conference, with the Johannesburg summit. On all these occasions, it was argued that to meet sustainable development goals, current trends of production, consumption, and distribution need to be halted.

8.6 Partnerships for sustainable globalization

The global economy demands adjustments that increase competitiveness and advance technological change to enhance productivity. Consultative and participatory approaches are needed to develop and implement effective socio-economic policies at the national and global levels. The poverty effects of trade liberalization are transmitted via four broad groups of institutions (whether net suppliers or end users of the goods or services in question): enterprises, distribution channels, the government, and households. Policies promoting decent employment opportunities, complemented by social protection and well-functioning labor markets institutions, are crucial.

Many people are endowed with specific capital, i.e., skills or equipment needed by industry. As industries and markets change, people gain or lose along with their employers—and can find the basis for their livelihood destroyed. Public policy can only partially acknowledge and mitigate such costs that can continue for months (or even years), as unemployed people search to relocate, retrain, reinvest, and otherwise readjust. Those later in life may never recover. Economists therefore usually favor only a gradual movement toward free trade, so that adjustment costs can be accommodated within the routine ups and downs of market dynamics (Deardorff and Stern, 2001).

Sustainable development requires mechanisms that permit the government and the private sector to develop "shared" vision, planning, and decision making. Economic planning and policymaking need to become more prudent and transparent, respecting the interests of future generations. This calls for a new type of partnership that redefines the roles and responsibilities of public and private sector organizations. Employers' and workers' organizations may find that they should review their roles and responsibilities and take a more central role in national policy and program cycles. Consultations at all levels will contribute to consensus-building, accountability, and transparency. Civil society representatives and other national and international partners, together with central and regional government, should be actively involved in addressing specific issues of local development. Such joint efforts will strengthen the three pillars of sustainable development:

- pursuing economic, social, and environmental objectives,
- linking sector strategies to macroeconomic concerns,
- ensuring policy coherence.

A better understanding of the causal mechanisms of globalization and consultation with social partners might contribute to the development of an early warning system and increase social risk management. In turn, this might enable countries to offset the transitional social costs of globalization and alleviate attendant increases in poverty. A broad participation of social partners is needed to bring key information into the policy implementation

Table 8.1 The roles of social partners for sustainable development.

Government	Employers' organizations	Trade unions
• Social protection • Employment strategies	• Define new business and restructuring plans • Labor force analysis identifying skill mix and new skill requirements	• Agreements
	Social dialogue to share the burden	

process. When implementing trade-related reforms, policymakers need to be aware of potential shocks—who would win and who would lose. All stakeholders need to adapt to the new situation. This can be done through a socially and environmentally responsible restructuring of the economy, modernizing the institutions and the industrial base, and building a national capacity to foster innovation. Flexibility should be increased by developing negotiating and administrative skills for crisis management (Essama-Nssah, 2000) (Table 8.1).

In response to the challenges of globalization, the private sector is more widely addressing its role in sustainable development as companies realize that it is in their commercial interest that the community be vibrant and prosperous. There is increasing evidence of private sector willingness to invest in social projects, as part of building success in their core business. Also, businesses are backing traditional philanthropic initiatives (World Bank, 1998).

Socially and environmentally responsible initiatives toward sustainable growth are essential and should be widely promoted. Table 8.1 outlines the main roles for each of the social partners. Within the tripartite collaboration, the Governments will take the lead for development of the Social Protecion and Emplyoment Strategies. Whereas the Employers' organizations will be actively engaged in defining new business plans and help with labour force analysis to identify the skill mixes and new skill requirements where necessary. Jointly with Trade Unions, tripartite Agreements will be forged and implemented.

8.7 Social assessment

In today's global world, national economies are integrated through production, distribution, and consumption chains—and a downturn in any country easily spills over to others. Recent financial and sanitary crises demonstrate the need to enhance social protection systems. Policies should concentrate on social security, as well as employment security. Mechanisms of social assessment (SA) are needed both for domestic policy interventions and to counteract external financial and economic shocks.

Article 22 of the Universal Declaration of Human Rights (UN, 1948) states that "everyone, as a member of society, has the right of social security." Article 9 of the International Covenant of Economic, Social, and Cultural Rights (UN, 1966) refers to the "right of everyone to social security, including social insurance."

The report of the secretary-general to the special session of the General Assembly on *Assessing the Social Impact of Economic Measures Taken in Response to Financial Crises in Developing Countries* (UN, 2000a) recognized that in countries undergoing structural adjustments and economic reforms, there is a need for regular and frequent monitoring and evaluation of the social impact of such changes and reforms. Affected groups of the population need to be better targeted so that emergency programs can be better designed. The report stresses that SA is particularly important for countries where representative democracy is weak. SA can serve to demonstrate the positive economic impact of social policies and programs (such as social protection, the identification of labor standards, investments in health, education, and basic social services) on the productivity of the workforce. In this way, it can contribute to increased output and improved competitiveness and profitability of enterprises.

Typical indicators of social impact measure employment, and absolute and relative income; levels of education, health, housing, and other social services; as well as social integration. The report stresses the importance of the timing and sequencing of policy measures so that their impact becomes positive, improving work conditions. This all depends on the effectiveness of investments in infrastructure and the development of social and human resources. SA can assist to:

- forecast and evaluate the impact of policies, programs, and projects; identify factors that improve the response of the poor to incentives for raising their productivity and income;
- analyze the poverty reduction and employment content of public investments;
- design social protection systems for the poor;
- strengthen the ownership of programs by participatory processes that involve their beneficiaries and other relevant actors.

Forecasting the social impact of policies and programs and evaluating them *ex post* has been a common practice for some time. The critical importance of SA was earlier underscored at the World Summit for Social Development convened by the UN Economic and Social Council (ECOSOC) in Copenhagen (UN, 1995). The program of action stated that an integrated socioeconomic approach to development requires, inter alia:

- *analyzing* and reviewing macroeconomic, microeconomic, and sectoral policies and their impact on poverty, employment, social integration, and social development;
- *assessing* the extent, distribution, and characteristics of poverty, unemployment, social tension, and social exclusion;
- *developing quantitative and qualitative indicators* of social development, poverty, employment, social integration, etc.

SA is usually considered to be an essential component of policy formulation and a tool to understand how the social welfare of different groups will be affected by development initiatives. Yet, some main questions of SA (World Bank, 1998) often remained unanswered:

- What will be the impact of globalization on various stakeholders particularly on women and vulnerable groups?
- Are there plans to mitigate adverse impacts?
- What social risks might affect project or program success?
- What institutional arrangements are needed for participation and project delivery?
- Are there adequate plans to build the capacity required at the appropriate level?
- How can existing tools and techniques be used to study the social impact of proposed reforms?
- Have alternative policy options been considered?
- How can policies be appropriately sequenced?

These questions all aim at a better understanding of the channels through which globalization and trade-related policies impact the income distribution, employment, and poverty.

CHAPTER NINE

Toward a Geneva Consensus

The economist as such does not advocate criteria of optimality. He may invent them. He will discuss their pros and cons, sometimes before but preferably after trying out their implications. He may also draw attention to situations where allover objectives, such as productive efficievncy, can be served in a decentralized manner by particularized criteria, such as profit f maximization. But the ultimate choice is made, usually only implicitly and not always consistently, by the procedures of decision making inherent in the institutions, laws and customs of society. A wide range of professional competences enters into the preparation and deliberation of these decisions. To the extent that the economist takes part in this decisive phase, he does so in a double role, as economist, and as a citizen of his polity: local polity, national polity, or world polity.

(Koopmans, 1975)

9.1 Sustainability and trade-offs

The presence of poverty among plenty, of need amidst unused resources, and destitution in the footpath of riches is not a new problem facing humanity. Yet, as stated in the recent report of the UN Secretary-General: *A Life of Dignity for All* (UN, 2013), the world's quest for dignity, peace, prosperity, justice, sustainability, and an end to poverty has reached an unprecedented moment of urgency.

The "Earth Summit," held in Rio in 1992, established sustainable development as the guiding vision for the development efforts of all countries. Governments also undertook to formulate and implement national sustainable development strategies. These were envisaged to be highly participatory instruments intended under the point 8.7. of the report of the United Nations Conference on Environment and Development "to ensure socially responsible economic development while protecting the resource base and the environment for the benefit of future generations" (UN, 1992).

In 2000 the member states of the United Nations reaffirmed the fundamental values of freedom, equality, solidarity, tolerance, and held up the bold vision of a future based on respect for the integrity of the planet and on shared responsibility. That vision, enshrined in the Millennium

Measuring Sustainable Development Goals Performance
ISBN: 978-0-323-90268-7
https://doi.org/10.1016/B978-0-323-90268-7.00011-6

Declaration (General Assembly resolution 55/2) and rooted in the Charter of the United Nations, recognizes the need to advance on three fronts simultaneously: development, peace and security, and human rights.

In 2001 the High-Level Meeting of the Organization for Economic Cooperation and Development (OECD), Development Assistance Committee (DAC) endorsed "Strategies for Sustainable Development: Practical Guidance for Development Cooperation," to provide policy guidance on good practice in developing and implementing strategies for sustainable development (OECD, 2001). Analyzing the international experience over the two decades from a process of multistakeholders dialog, in a number of developing countries, assessing their experience of country-level strategies for sustainable development, OECD draws important lessons that are of equal relevance not only to developing but also to developed countries alike.

While the main aim of the OECD Guidelines is to provide a framework for development cooperation agencies in their efforts to assist developing countries toward sustainable development, it could also serve other major development stakeholders: policy-makers, planners and development practitioners, as well as to academics, students, and development analysts in all countries.

What are strategies for sustainable development? OECD/DAC Guidelines on page 16 defines a strategy for sustainable development as comprising:

> A coordinated set of participatory and continuously improving process of analysis, debate, capacity-strengthening, planning and investment, which seeks to integrate the short and long term economic, social and environmental objectives of society- through mutually supportive approaches wherever possible- and manages trade offs where this is not possible.
>
> ***(OECD, 2001)***

In other words, the OECD acknowledged that "in simple terms, sustainable development means integrating the economic, social and environmental objectives of society, in order to maximize human well-being in the present without compromising the ability of future generations to meet their needs. This requires seeking mutually supportive approaches whenever possible, and making trade-offs where necessary." Within the above context, coordinate implies negotiation of trade-offs and conflict management.

Since Rio Summit, international community has inspired to integrate the social, economic, and environmental dimensions within a single framework. Nonetheless, no country has yet succeeded. Prevailing governance systems continue to separate these pillars at policy, planning, and management levels. There is an accumulating evidence that this disconnect has led to a widening of inequality, both at global and national levels, pervasive informalization and social tension, alarming pace of climate change, and environmental degradation.

And there is a strong consensus among all development practitioners, that to meet the challenges of sustainable development, strategic planning practices need to become more effective, efficient, credible and lasting, a standardized or blueprint approach is to be avoided, and institutional arrangements and procedures according to individual countries' own needs, priorities, and resources need to be established.

It is well known that good strategies for sustainable development entail the integration and balancing of the above-mentioned factors, where possible, and making choices and trade-offs, where necessary. However, past strategies have not effectively dealt with the trade-offs, even when sustainable development and poverty reduction is declared to be the objective of country development strategies because the methods remain poorly developed and the necessary skills and/or political will have been lacking.

There is a strong consensus that although economic, social, and environmental objectives are often complementary, there is no natural trade-off between them and that it is a matter of policy choices and prioritization in the framework of given strategic development goals of countries. The choice among these objectives inevitably leads to stakeholders' conflicts. Addressing it without compromising our shared prosperity and the right to development of future generations requires a significant alteration of macroeconomic governance and calls for innovative policy responses toward the achievement of sustainable development goals (SDGs).

However, these legitimate but opposing interests and objectives cannot always be balanced by judicial systems. The need for social partnership and dialogs that can facilitate the process of negotiating compromises between stakeholders has emerged. This requires participation in decision-making, genuine dialog and, most importantly, a strong political will, since it is impossible to separate politics from economics.

Ensuring participatory approaches and active participation of all stakeholders enables us to consider the data, insights, and information during the analysis that is crucial to the outcome of a successful, integrated assessment.

The complex relationship between economic and social development makes it really difficult to establish trade-offs. This is one of the main problems at any level as it is not easy to convince decision makers to give higher priority to short-term social objectives than long-term economic interests, especially where financial resources are scarce. The choice between these interests and objectives inevitably leads to conflicts between stakeholders.

Making globalization serve human development and ensure sustainable improvements calls for careful timing, speed and sequencing of trade/investment liberalization reforms, and a need to make trade-offs, where necessary. To overcome the contemporary challenges of globalization, a strong political commitment and institutional capacity will be needed.

9.2 What should economists do?

The declaration of the Rights of Man and of the Citizen, dating back to the French revolution, served as a founding stone for universal human rights. It is valid at all times and in every place and it pertains to human nature itself. Transferring the concept of liberty to the sphere of economics, the idea of economic liberalization can only be ethically and politically motivated provided that such liberty does not injure others (or at least minimizes possible negative impacts on others), that it applies to all members of society (or, failing this, that timely compensatory mechanisms are established), and that the limits of economic liberalization are defined by law or existing social and physical infrastructure. As a consequence, even a liberal economic regime needs institutional support and state regulations.

Economists have long been in a search of a theory that would enable the society to make trade-offs leading to social utility maximizing decisions. James Buchanan is widely perceived as an iconic figure in political economy—for his passion in analytical and ideological innovations, and as the champion of the methodological individualist perspective across the social sciences. He was much immersed in exploring the questions associated with a society of free and responsible individuals and identification of rules that would enable the society to live better together while simultaneously respecting individual autonomy and generating economic prosperity for all.

The persistent and consistent application of opportunity cost reasoning, the treatment of the status quo, the analysis of social change, and the deployment of the compensation principles had long occupied his mind. In 1964, challenging the conventional wisdom he boldly asked a question: "**What Should Economists Do?**"

In his attempt to answer the question, Buchanan, reverted back to, what he considered to be, a much-neglected principle—"to truck, barter, and exchange," which gives rise to the division of labor, enunciated by Adam Smith. In Chapter Two of "The Wealth of Nations":

> It is not originally the effects of any human wisdom, which foresees and intends that general opulence to which it gives occasion. It is the necessary, though very slow and gradual, consequence of a certain propensity in human nature which has in view no such extensive utility; the propensity to truck, barter, and exchange one thing for another.

With that in mind, he concluded that the economist "should" concentrate their attention on a particular form of human activity, and upon the various arrangements that arise as a form of activity.

Buchanan refereed critically to "The Nature and Significance of Economic Science" of Lord Robbins and his formulation of the economic problem, as the allocation of scarce means among alternative or competing ends necessitated by the fact of scarcity and a need to choose and making of allocative decisions. Taking Robbins as an adversary, he stated categorically, that the statement of Robins, "his all-too- persuasive delineation of our subject field," that dominated economics textbooks, "has served to retard, rather than to advance scientific progress."

Recalling, the presidential address Howard S. Ellis delivered to the American Economic Association in 1949, he agreed that:

> it is thus by quite natural or normal extension that the economic problem moves from that one which is confronted by the individual person to that facing the larger family group, the business firm, the trade union, the trade association, the church, the local community, the regional or state government, the national government, and, finally, the world.

A this critical congestion point in global decision-making to implement the Agenda 2030, that entail consideration of conflicting interests, I could not agree more with Buchenan, in *want*[ing] *economists to modify their thought processes, to look at the same phenomena through* "another window," or as he puts using Nietzsche's appropriate metaphor, *I want them to concentrate on* "exchange" *rather than on* "choice."

In a search for answer for the whole of the relevant collective group of individuals, who have organized themselves socially for solving their economic problem, qualified as a society, expanding to "economic system" and "social organization," he further agreed with the principles of political-philosophical order suggested by Frank Knight and Milton

Freidman and their identification of the "society" as the entity that confronts the economic problems, whose ends are to count in the appropriate calculus of margins. Preferring the definition of economics by the latter, as "the study of how a particular society solves its economic problem," Buchanan acknowledges that there is meaningful content in economics for "social welfare."

Buchanan's reference to the scientific vocabulary used by a British philosopher Thomas Weldon, in that, the very word "problem" in and of itself implies the presence of "solution" is of particular importance in addressing the contemporary challenges of attaining sustainable globalization. And, I should join him and Rutledge Vining:

> Our whole study becomes one of applied maximization of a relatively simple computational sort. Once the **ends to be maximized** are provided by the social welfare function, everything becomes computational .... If there is really nothing more to economics than this, we had as well turn it all over to the applied mathematicians.

Introducing a notion of "social engineering," Buchanan makes an attempt to distinguish between the economic and the technological problem—while in the latter "the function to be maximized is specified in advance of the choices to be made" it is not the case in the former! Referring to the paradox entailed in the theory of choice, he states:

- "If the utility function of the choosing agent is fully defined in advance, choice becomes purely mechanical. No "decision," as such, is required; there is no weighing of alternatives. On the other hand, if the utility function is not wholly defined, choice becomes real, and decisions become unpredictable mental events. If I know what I want, a computer can make all of my choices for me. If I do not know what I want, no possible computer can derive my utility function since it does not really exist. But the distinction to be drawn here is surely that about the knowledge of the utility function."
- "An economic problem arises when mutually conflicting ends are present, when choices must be made among them. A technological problem, by comparison, is characterized by the fact that there is only one end to be maximized."

The attainment of sustainable development and simultaneous pursuit of multiple development goals, and the political pressure of making trade-offs between 17 SDGs, and the consensus on forging global partnership for development, calls for collaboration, coordination, and coherence among all development stakeholders and connects well with the approached advocated by Buchanan:

(i) recognition that at each stage in this evolution toward solution, there are mutual gains to be made and they can be secured through cooperative endeavor, through "exchange," "trade," or "agreement,"

(ii) acknowledgment that the mutuality of advantage that may be secured by different organisms as a result of cooperative arrangements, be these simple or complex, shall make the target of the discipline.

Within that context, his definition of the notion of "Symbiotics" gains particular importance:

> as the study of the association between dissimilar organisms, and the connotation of the term is that the association is mutually **beneficial to all parties.** This conveys, more or less precisely, the idea that should be central to our discipline. It draws attention to a unique sort of relationship, that which involves the co- operative association of individuals, one with another, even when individual interests are different.

Comparing the "market," as a mechanism with "government," and as an alternative mechanism for accomplishing similar tasks, Buchanan describes "market" as an engineered construction, a "mechanism," an "analogue calculating machine," a "computational device," one *that processes information, accepts inputs, and transforms these into outputs which it then distributes.* He then defies "efficiency"—the core motivation for individuals to engage in trade, the source of the propensity, *as the personal sense of moving from less preferred to more preferred positions, and doing so under mutually acceptable terms.*

He further recognizes that, it is contradictory to talk of the market as achieving "national goals," efficiently or inefficiently, and an "inefficient" *institution, one that produces largely* "inefficient" *results, cannot, by the nature of man, survive until and unless coercion is introduced to prevent the emergence of alternative arrangements.*

Buchanan draws distinction between economics and politics, linking them to the nature of the social relationships among individuals that is examined in each disciplines. While the predominant characteristic of "economic" behavior involves individuals exchanging, trading, as freely contracting units, whereas going beyond the ordinary price-money nexus, the "political" behavior is characterized by the nature of a relationship between individuals when meeting one another: superior-inferior, leader to follower, principal to agent.

Stressing the presence of coercive or quasi-coercive elements in decision-making in the sociopolitical institutions, Buchanan emphasized, that in the political deliberations a man is empowered with an "equal," or a trading relationship with his associates to the extent that he has available alternatives of action. He also clarified that the economic relationship is wholly replaced by the political only in those situations where pure rent is the sole element in return.

Buchanan stated that the advances in our understanding of symbiotic relationships may well require considerably more sophisticated mathematical tools than those required "social engineering" leading to the development of more and more complicated schemata designed to maximize more and more complex functions, under more and more specifically defined constraints. He identified a need to learn much more about the *theory of n-person cooperative games*.

> It seems but natural that the mathematics finally required to systematize a set of relationships involving voluntary behavior on the part of many persons will be more complicated than that required to solve even the most complex computational problem where the ends are ordered in a single function.

In 2004 Finn Kydland during his Novel Lecture[1] recalled the words of Robert Lucas, 1980 Nobel Laurate in Economics, who defined the functions of theoretical economics:

> to provide fully articulated, artificial economic systems that can serve as laboratories in which policies that would be prohibitively expensive to experiment with in actual economies can be tested out at much lower cost ... Our task, as I see it ... is to write a FORTRAN program that will accept specific economic policy rules as 'input' and will generate as 'output' statistics describing the operating characteristics of time series we care about, which are predicted to result from these policies[2]

Kydland reflected on the desired environments referred by Lucas suggesting to place *people in desired model environments* and record how they behave under alternative policy rules and rightly stated that the required information on "individual responses could be documented relatively cheaply by means of ... censuses, panels [and] other surveys ..."

[1] Finn E.K. Prize Lecture. NobelPrize.org. Nobel Prize Outreach AB 2021. October 19, 2021, https://www.nobelprize.org/prizes/economic-sciences/2004/kydland/lecture/.

[2] Robert E.L., Jr. Prize Lecture. NobelPrize.org. Nobel Prize Outreach AB 2021. October 18, 2021, https://www.nobelprize.org/prizes/economic-sciences/1995/lucas/lecture/

Another Nobel laureate Ronald Coase in his Remarks at the University of Missouri Columbia, Missouri, on April 4, 2002 entitled "Why Economics Will Change" (Coase, 2002) had criticized this failure to look at the system as the object of the study and highlighted objectively that it may well take quiet long time to do so:

> Now how does it look? Well, that it's complicated is not a question, nor that it could take a long time if we start, as I hope we will, to study this system. Studying it may take one hundred years. It may take two hundred years. But anyway, we should start.

To deliver on the "Future We Want" for our future generations we should start now!

9.3 Bridging the gap between science and policy

In his seminal book *The Nerves of Government* Karl Deutsch (1963) raised the issue of the optimistic or a conservative bias implicit in the notion of a self-steering society. He compared the process of steering a ship and the governing of human organizations:

> Steering a sheep implies guiding the future behavior of the ship on the basis of information concerning the past performance and present position of this ship itself in relation to some external course, goal, or target. In such cases, the next step in the behavior of the system must be guided in part by information concerning its own performance in the past.
>
> ***(Deutsch, 1966)***

There are lessons to be drawn from this analogy. The current model of globalization is clearly yielding uneven results for the sustainable development outcomes of nations. Not all nations succeed in reaping the fruits of modern technology or making globalization serve their national interest. They find it hard to create "shared prosperity"[3]—the only kind of prosperity that can truly be sustainable. The new post-2015 development agenda calls for the integration of the multiple dimensions of sustainable development into a single framework, relying on the evaluation of past successes and failures.

[3] https://www.rawstory.com/2013/09/nobel-laureate-stiglitz-the-only-true-and-sustainable-prosperity-is-shared-prosperity/.

The *Oxford Handbook on Governance* (Levi-Faur, 2012):

> outlining formalized models of governance highlights the importance of information gathering and information processing in governance. It acknowledges that information will often be politicized prior to being acted upon. Someone must consider the governance environment, react to information collected, make decisions and then continue to consider the consequences of each round of governing.

A Report of the Commission of Experts of the President of the UN General Assembly on Reforms of the International Monetary and Financial System published in September 21, 2009 stressed:

> Reform of the international system must have, as its goal, the improved functioning of the world's economic system in support of the global good. This entails simultaneously pursuing long term objectives, such as sustainable and equitable growth, creation of employment in accordance with the "decent work" concept, responsible use of natural resources, reduction of greenhouse gas emissions, and more immediate concerns, including addressing the challenges posed by the food and financial crises and global poverty.[4]

The report noted that the crises, as well as the growing divide between poor and rich within and between countries and the risk of systemic climate change, are all interconnected global challenges that threaten to unravel the fragile state of globalization. It called for UN reform recognizing these overarching priorities and to deliver on its longstanding commitments. Recognizing that "for too long we have sought to burn and consume our way to prosperity" and that the UN must use science and technology to strengthen its policy-making, the Secretary-General Ban Ki-moon on 30 January 2015 launched an advisory board of 26 eminent scientists. The board aims at strengthening the ties between the UN and the scientific community so that science can be more effectively integrated in the policy-making processes.

The Secretary-general's high-level panel on the post-2015 Development Agenda reiterates the vital importance of building data systems that can provide timely, disaggregated indicators to measure progress in all countries, and at all levels (local, subnational, and national). Toward this end, the report "A New Global Partnership: Eradicate Poverty and Transform Economies through Sustainable Development,"[5] released in 2013 recommended

[4] http://www.un.org/ga/econcrisissummit/docs/FinalReport_CoE.pdf.

[5] United Nations, 2013. A New Global Partnership Eradicate Poverty And Transform Economies Through Sustainable Development - The Report of the High Level Panel of Eminent Persons on The Post-2015 Development Agenda, Report of the High-level Panel of Eminent Persons on the Post-2015 Development Agenda. (United Nations: New York).

establishing a Global Partnership on Development Data that will bring together diverse but interested stakeholders—government statistical offices, international organizations, CSOs, foundations, and the private sector.

Throughout my career, I have had the unique opportunity of acting both as an economist and a politician, at national and international levels, in terms of definition of Buchanan, whereby:

– *Economics is the study of the whole system of exchange relationships,*
– *Politics is the study of the whole system of coercive or potentially coercive relationships.*

9.4 Rating country performance by frontier analysis

Spurred on by the success of the joint European Commission, European Parliament, Club of Rome, OECD, and World Wildlife Fund high-level conference entitled "Beyond GDP" held in 2007; the Beyond GDP initiative is about developing indicators that are as clear and appealing as GDP, but more inclusive of environmental and social aspects of progress. Economic indicators such as GDP were never designed to be comprehensive measures of prosperity and well-being. We need adequate indicators to address global challenges of the 21st century such as climate change, poverty, resource depletion, health, and quality of life.

The Synthesis Report of the Secretary-General of the Post-2015 Agenda "The Road to Dignity by 2030" (§ 133) states "Progress in sustainable development will depend on vibrant economies and inclusive growth to keep pace with growing populations and longer life expectancies, and to generate employment, wages, and revenues for social programmes. But for making our economies inclusive and sustainable, our understanding of economic performance, and our metrics for gauging it, must be broader, deeper and more precise."

Recalling earlier papers by Dan Usher, "The Economic Prerequisites to Democracy," Rutledge Vining, "On Appraising the Performance of an Economic System," and, a paper he wrote with Geoffrey Brennan, "The Reason of Rules," Buchanan acknowledged that:

> it is not legitimate to criticize, for example, an existing distribution of income of allocation of resources as being un just, inequitable, or inefficient, without being able, at the same time, to demonstrate some proposed alternative regime that can be expected to generate distributions or allocations that will do better by the same standards.

As already presented in earlier part of the book, Data envelopment analysis (DEA), co-developed by Abraham Charnes, William Cooper and Edwardo Rhodes in 1978, is ideally suited for the purpose, delivering a metric that evaluates the performance of decision and policy-making units (DMUs/PMUs) (nations, regions, cities, enterprises, banks, schools, etc.) according to their ability to use inputs, and marshal economic and social policy to attain a spectrum of desired goals.

In our earlier work, we have developed an innovative framework to measure the relative socioeconomic performance of countries in the context of globalization (see Adler et al., 2010; Thore, 2014; Thore et al., 2015; Thore and Tarverdyan, 2008; Thore and Tarverdyan, 2009; Cooper, Thore and Tarverdyan, 2010; Thore and Tarverdyan, 2015; Thore and Tarverdyan, 2016). Each paper used different set of countries, policy and performance variables, and answered to a different question using a specific formulation of a DEA model.

In Thore and Tarverdyan (2008) we used the DEA to screen the possibility of a Fair Globalization. We estimated a social preference function of the various dimensions of Decent Work for 72 countries from all continents. The calculations took the form of estimating a generalized input-output relationship, employing Decent Work as the outputs and a set of factors measuring the effects of globalization and economic and social policy parameters as inputs. An efficiency frontier of "decent" work, incorporating aspects of employment, absence of poverty, and the ratio of women in the labor force was evaluated. The "decent" work deficit for each inefficient nation was calculated, given the level of economic globalization, trade, and capital liberalization. In this simulation, we solved the output-oriented DEA program with flexible returns to scale.

In a subsequent piece of work Thore and Tarverdyan (2009) we made an attempt to quantify the International Labour Organization (ILO) objectives and identify policies conducive to Decent Work in the globalized world. A related data base for 61 countries was used to estimate "A Fair Globalization" as another social preference function, blending aspects of the open and unrestricted economy with measures of social protection. We again estimating the frontier of a generalized input-output function, using the dimensions of a fair globalization as outputs. The inputs listed a set of factors measuring the degree of receptivity of the winds of globalization of each country. An idealized efficiency frontier with respect to fair globalization was formed whereby the efficiency frontier traces the maximal level of trade and social protection that the current level of globalization suggests is attainable. In this second simulation, we solved the output-oriented DEA program with flexible returns to scale.

In two studies, Thore and Tarverdyan defined two connected efficiency concepts, each defining a trade-off between economic and financial variables on the one hand, and social and policy variables on the other.

In Adler et al. (2010), we developed a framework to estimate the relative efficiency of developing countries in utilizing both their domestic and external resources to achieve the Millennium Development Goals. Our analysis highlighted distinct efficiency differences across lending groups and geographic regions, e.g., between Sub-Saharan Africa and Latin America, the 61 countries.

The performance of regions exposed to natural disasters and political violence need to be analyzed individually and risk management ought to be an integrated part of development policy. The gap between the Millennium Development Goals and human rights approaches is then examined. Finally, we qualify a country's performance level as potentially sustainable in terms of human, social, and environmental welfare, in turn searching for realistic benchmarks and intermediate targets for the relatively inefficient countries.

We have applied principal component analysis in combination with DEA (PCA-DEA) to solve the problem of efficiency overestimation through data aggregation. Using PCA-DEA, we solved the multidimensional problem of relative efficiency estimation and benchmarking, using small space analysis to present the issues graphically. Our aim was not to rank countries in a league table rather to provide a framework that combines economic, environmental, and social issues in order to search for sustainable, pragmatic benchmarks, pushing the boundaries of the Human Development Index.

The applied basic DEA model may not sufficiently distinguish between "efficient" and "inefficient" decision-making units (DMUs) due to overestimation bias and sometimes more than half of the DMUs lie on the frontier, categorizing them all as efficient. This result is directly due to the curse of dimensionality caused by a relatively large number of variables in comparison to the number of observations. To deal with this shortcoming, Adler and Golany developed a methodology to reduce the number of inefficient DMU's incorrectly defined as efficient, is PCA-DEA (Adler and Golany).

PCA is applied to replace the original outputs with a smaller group of principle components (PCs), which explain the variance structure of a matrix of data through linear combinations of variables. As stated in

Hair et al., let the random vector $Y = [Y_1, Y_2, ..., Y_p]$ (i.e., the original outputs chosen to be aggregated) have the correlation matrix C with eigenvalues $\lambda_1 \geq \lambda_2 \geq ... \geq \lambda_p \geq 0$ and normalized eigenvectors l_1, l_2, ...,l_p. Consider the linear combinations, where the superscript t represents the transpose operator:

$$
\begin{aligned}
&Y_{PC_i} = l_i^t Y = l_{1i} Y_1 + l_{2i} Y_2 + ... + l_{pi} Y_p \\
&\mathrm{Var}(Y_{PC_i}) = l_i^t C l_i, \qquad i = 1, 2, ..., p \\
&\mathrm{Cov}(Y_{PC_i}, Y_{PC_k}) = l_i^t C l_k \ , \ i = 1, 2, ..., p, \ k = 1, 2, ..., p, \ i \neq k
\end{aligned}
\tag{9.1}
$$

The principal components are the uncorrelated linear combinations $Y_{PC_1}, Y_{PC_2}, ..., Y_{PC_p}$ ranked by their variances in descending order. In order to apply PCA-DEA, each original output is divided by the corresponding standard deviation and the correlation matrix of standardized outputs and principal components are calculated. It is noteworthy that principal components are based solely on the correlation matrix and their development does not require a multivariate normal assumption. In order to use PCs instead of the original data, the DEA model needs to be transformed to take into account the linear aggregation.

The PCA-DEA formulation for the radial, output-oriented, variable returns to scale linear program applied in our study is presented in Model (9.1), where subscript *'o'* (*'pc'*) is the index of original (PC) variables; Y_{pc} represents principal components based on the output matrix; Y_{pc}^a output column vectors for DMU a, respectively; λ a column n-vector of DMU weights; σ a column m-vector of input excess variables; s a column r-vector of output slack variables; e^t a row vector of ones, and V^t and U^t input and output row vector weights, respectively. L is the matrix of all normalized eigenvectors whose dimensions drop as PCs are removed from the analysis.

$$
\begin{aligned}
&\underset{s,\sigma,\lambda,\theta}{\mathrm{Max}\,\theta} \\
&s.t. Y_{pc}\lambda - L\,s = \theta Y_{pc}^a \\
&- X_o\lambda - \sigma = -X_o^a \\
&e^t\,\lambda = 1 \\
&\sigma \geq 0 \\
&s \geq 0 \\
&\lambda \geq 0
\end{aligned}
\tag{9.2a}
$$

$$\begin{aligned}
&\underset{V,U}{\text{Min}}\, V^t X_o^a + U_o \\
&s.t. \quad U^t Y_{pc}^a = 1 \\
&\qquad V^t X_o - U^t Y_{pc} - U_o \geq 0 \\
&\qquad V^t \geq 0 \\
&\qquad U^t L \geq 0 \\
&\qquad U_o \text{ and } U \text{ are free}
\end{aligned} \tag{9.2b}$$

Model (9.1) ensures that were we to use all PCs, we would attain exactly the same solution as that achieved under the original DEA formulation.

It enables an analysis of the current level of multiple dimensional development of a nation, achieved through the optimal use of available resources and can be used for the allocation of official development assistance (ODA) and other financial resources effectively and efficiently in order to support positive future trends toward meeting SDG targets. The derived scores are based on the estimates of relative values rather than a priori imposition.

In Cooper et al. (2010) dealing with situations of economic and social planning, the same approach was used to determine the relative effectiveness of the policy achievements. Its basic premise was the calculation of a measure of effectiveness, presented as a generalization of the engineering concept of the effectiveness of a machine. Effectiveness is a ratio of cardinal numbers, measuring the cardinal value of outputs per unit of inputs. In that vein, we calculated the effectiveness of labor market policy in terms of the goals of decent work. Similarly, we calculated the effectiveness of globalization in terms of a set of goals of social fairness. To facilitate comparisons with the two earlier works Thore and Tarverdyan (2008) and (2009), we made the calculations for a list of 52 countries. The list was the intersection of those two earlier lists covering countries that were featured in both of these earlier studies.

An aggregate of outputs was formed—the social welfare V(Y) of the goals. Similarly, an aggregate of inputs was formed—the function U(X) of the policy parameters. We have already defined (see Eq. 1.5) the effectiveness rating of a given policy-making unit (PMU), such as a country, as the ratio between the social welfare obtained V(Y) and the utility cost of the policies employed U(X). The set of calculations now indicated aim at maximizing the effectiveness ratio V(Y)/U(X) while seeing to it that the effectiveness measures of all other PMUs remain meaningful, i.e., that they all be less than or equal to 1. For the PMU currently to be rated (say $j = 0$), the maximization problem would then read using the logarithmic specifications U and V (see Eq. 1.6).

The output-oriented CCR with the given observations replaced by their logarithms was applied. All statements being in logarithmic units suggests that we might consider a multiplicative model as a generalization of the Cobb Douglas (log linear) function. See DEA page 110 in the text by Cooper, Seiford Tone for references and Section 6 of Chapter 2 in the Handbook on DEA (Kluwer Academic Publishers) for use for returns to scale.

Our last paper Thore and Tarverdyan (2016), entitled "The Sustainable Competitiveness of Nations", we have estimated the multidimensional competitiveness for 82 countries from all continents using the output-oriented CCR with the given observations. Using 12 pillars of the World Economic Forum as Inputs and 8 sustainability goals' indicators, we recalculate the sustainable competitiveness index (SCI) defined by the Forum.

This last study has had the highest policy impact so far. The opinion of the European Economic and Social Committee (C228/37) on "Listening to the citizens of Europe for a sustainable future (Sibiu and beyond)" borrowed our approach and defined "sustainable competitiveness" as a model that balances economic prosperity, environmental issues, and social inclusiveness. EESC submits that, in this context, the sustainability-adjusted global competitiveness index (GCI) needs to take into consideration two new dimensions—environmental and social. This definition is based on our DEA approach on evaluating the sustainable competitiveness of nations. It is noteworthy that the EESC preferred our definition to the World Economic Forum's with a majority of votes: For: 157; against: 64, and abstentions: 13.

Through its flagship Global Competitiveness Report, the World Economic Forum has over the course of the last five decades studied and measured the drivers of national competitiveness including a wide range of factors such as governance, infrastructure, education, innovation, and the proper functioning of markets.

To measure the impact of sustainability on economic competitiveness, in October 2011, the Forum initiated the development of a new analytical framework—the SCI; submitting that "Fundamental to the concept of sustainable competitiveness is the notion that, although competitiveness can be equated with productivity, sustainable competitiveness can be linked to a broader concept that focuses on aspects that go beyond mere economic outcomes. They define "**sustainable competitiveness**" *as the set of institutions, policies, and factors that make a nation productive over the longer term while ensuring social and environmental sustainability.*

The two areas of sustainability—social and environmental—are treated as independent adjustments to each country's performance in the GCI. The SCI widened the calculation of its standard index to include characteristics of demographics, social cohesion, and environmental stewardship. It was published in 2013–15 and 2015–15 and discontinued thereafter. The Forum acknowledged, that "the selection of their methodology is not intended to be scientific, but it represents a normative approach aimed at stimulating discussions on policy priorities and scientific research in this field."

Breaking with the precedent of the Forum, however, we have argued against calculating a single index, instead advocating frontier methods. In our empirical work, we have chosen four of its social factors: *the Gini index, youth unemployment, access to sanitation, access to improved drinking water* and four environmental factors: *agricultural water intensity, CO_2 intensity, the overexploitation of fish stocks, forest cover change* as inputs.

If one wants to, one may consider this function as an "index." The dramatic difference here, however, is that the weight of each goal in forming the index is not taken as a given constant but rather as an unknown, to be determined by fractional programming—calculating the *optimal* weights. Our procedure draws on the techniques of frontier analysis (see Charnes, 1952; Charnes et al., 1994; Cooper et al., 2004a; Cooper et al., 2005; Cooper et al., 2006). Rather than nominating winners and losers in some quest for an ever-growing GDP, one instead looks for the nations that are able to convert their GDP most *effectively* into the SDGs. These are the "frontier" nations, defining the frontier of the doable. The others are sub-frontier, falling short of the doable.

The EESC is a unique forum for consultation, dialog, and consensus between representatives from all the different sectors of "organized civil society" including employers, trade unions, and groups such as professional and community associations, youth organizations, women's groups, consumers, environmental campaigners, and many more. Set up in 1957, the EESC has always been an integral part of the EU; advising EU decision makers and ensuring that ordinary people have a say in Europe's development. It aims to help ensure EU policy-making and legislation is more democratic, more effective, and genuinely reflects the needs of EU citizens.

The European Parliament, Council and Commission are legally obliged to consult the EESC when passing new laws. The EESC examines the proposals and draws up opinions based on agreement reached between its 350 members from 28 EU Member States. The opinions are debated and adopted by simple majority at EESC plenary sessions, and then sent to the EU lawmakers to contribute to the law-making process by influencing their thinking. I genuinely believe by the adoption of our definition by EESC on March 20, 2019, during its Plenary session No 542 with a majority of votes 217/6/8, we have made our modest contribution to the paradigm shift beyond economic toward sustainable competitiveness that will still yields its results in the years to come!

9.5 Data envelopment analysis dual framework, utility maximization, and exchange

James M. Buchanan, widely perceived as positive or normative economist, has made major contributions to the development of the contractual and constitutional bases for the theory of political decision-making and public economics. He was awarded the Nobel Prize in 1986 as the modern developer of the theory of public choice.

In his contribution to "The State of Economic Science: Views of Six Nobel Laureates," Buchanan called for a revival of "catallactics" (or "catallaxy") as the core of our discipline, and viewed economics, as a social science, claiming it should be about trade, exchange, and the many and varied institutional forms that implement and facilitate trade, including all of the complexities of modern contracts as well as the whole realm of collective agreement on the constitutional rules of political society. Boldly negative about the state of macroeconomic science, Buchanan presented his distinction between the utility maximizing and the exchange paradigms. Viewing "the economy as an engine, mechanism, or means, organized for the achievement of specifically defined purposes," he made few critical statements that are at the heart of our research and methodological achievements:

- *The observing political economist is unable, even conceptually, to construct a* "social welfare function" *that will allow him to carry out a maximization exercise.*
- *An economy that persistently generates wide swings in levels of income and employment would, appropriately, be deemed to be a structural failure, and such a pattern of results should offer incentives to investigate, locate, and identify the structural sources of the problem, leading ultimately to structural-institutional reform.*
- *The economy-as-order interpretation depends critically upon agreement as the criterion for assessment. Since there are no independently existent scalars, the only indication that traders have improved their position lies in their observed agreement. A positive welfare assessment becomes possible because the agreement has*

signaled mutually preferred change. Agreement is the means of defining Pareto (Wicksell) superiority, and it is the only means that exists.

- *My concern here, however, is with the set of constraints that are subject to deliberative change, and, hence, to choice. Because these constraints are general and extend over all participants in the political economy, any choice must be, by definition, public, in the classic public good sense of this term. A shift in constraints for any one actor must apply for all actors.*

Buchanan recalled the Robbin's (1932) definition of "the economic problem": as allocation of scarce resources among alternative ends. Due to the inability to meet all demands, scarcity, entails designation of a criterion for *better or worse* to assess the alternatives when making the choices.

- *This criterion emerges as some common denominator that allows the differing demands to be translated into a single dimension, which we then label as* "**utility**" *or* "**value**."
- *Scarce resources are allocated among alternative uses so as to secure maximum value when a unit of each scarce resource yields equivalent value in each use to which it is put.*
- *Satisfying this norm maximizes value subject to the resource scarcity constraints.*
- *Economics, as a realm for scientific inquiry, does indeed seem to be reducible to applied maximization; the calculus seems surely to be its basic mathematics.*

Asserting that game theory offers the appropriate mathematical framework that facilitates an abstract understanding of economics. Buchanan specified distinct difference between the *exchange* process and the *choosing* process: whereas the former necessary implies an interaction between (among) separate actors (participants), no one of which can choose among "solutions," in the latter, each participant does, of course, make choices among alternative bids and offers (strategies). *A solution to an exchange emerges only from the choices made, separately and independently, by all participants in the process.*

- *In exchange, as in ordinary games, players or participants may be modeled as behaving so as to maximize their separately defined utilities, subject to the constraints separately faced, as defined by the rules, the endowments, and the predicted responses of other participants. The standard maximizing behavior embodied in rational choice models may, of course, be accepted for this analytical exercise. But, in exchange, again as in ordinary games, neither any single player-participant nor the set of players-participants, as a group, treats the outcome of the process as a maximand.*

Since its inception in 1978, the DEA has immediately been recognized as a modern tool for performance measurement. After the introduction of the first models by Charnes et al. (1978), as a tool for measuring efficiency and productivity of DMUs. A large number of books and articles, accounting for about 12,000, representing significant breakthroughs in theory and applications, both public and private sectors, to assess the efficiency, effectiveness, equity, and productivity of their activities has been published. Several bibliographic collections report a comprehensive analysis and an extensive listing of DEA-related publications covering its four decades of history of theoretical and methodological developments from 1978 to end of 2016. Emrouznejad and Yang (2018), Emrouznejad and Thanassoulis (1996a,b, 1997), (Kharas et al., 2019), (Kroll, 2015), (Lafortune et al., 2020), (Seiford, 1997), (Schmidt-Traub et al., 2019), (SDSN TReNDS, 2020), Tavares 2002, (Adams and Judd, 2018), Gattoufi et al. 2004, (OECD/JRC, 2008), (Papadimitriou et al., 2019), and Emrouznejad et al. 2018, (Inter-Agency and Expert Group on Sustainable Development Goal Indicators, 2018).

The largest number of 691 articles were published in the European Journal of Operational Research that makes up to 23.23%, followed by 281 in the Journal of the Operational Research Society-9.45%, 255 in the Journal of Productivity Analysis-8.57%, 237 in the Omega (United Kingdom)-7.97%, 181 in the Expert Systems with Applications-6.09%, and 177 Annals of Operations Research-5.95%. See. Emrouznejad and Yang (2018). Agriculture, banking, supply chain, transportation, as well as public policy are the top five application fields of DEA with the greatest numbers of journal articles in 2015 and 2016. The *Journal of Socio-Economic Planning Sciences* has been recognized as the first choice journal for DEA papers with a great variety of applications in the public sector.

Intrigued by the considerations presented by the six Nobel Laureates, in the earlier pages of this book and awakened by the strong critics about the how badly economists have done their job in the past, especially stressed by Simon and Buchanan, we came *to* realize, that, in fact, the concepts embodied in two different definitions: *engineering-science* and *Pareto-Koopmans* efficiency and joined together by DEA dual models provide a "new" theoretical grounds for ensuring the long desired analytical underpinnings that the decision makers need for unification of the Buchanan's *utility maximizing* and the *exchange* paradigms into a single decision-making framework. While primal DEA Model will provide solutions guiding toward utility maximization, the dual variable (shadow price) derived from the dual models will help shape the discussion to make choices on possible trade-offs: *exchange involves agreement on the part of traders, both upon entry into trade and upon terms of trade ... If exchange moves each trader higher on the scalar assigned to him, the change is defined to have been Pareto superior.*

Having successfully constructed the "social welfare function," much inspired by many economists, this revealed unique potential of the DEA, holds a capacity to serve the expectations of Buchanan, We, therefore, suggest the DEA to become a strategic means of participatory decision-making process, encompassing the two dimensions of his *utility maximizing* and the *exchange paradigms* that will treat effectiveness-efficiency-equity trade-offs and will enable all development stakeholders to make informed choices and build a consensus on managing the process of globalization and plan future development policies toward the achievement of SDGs in different contexts and different levels.

Based on the main DEA models, presented in Cooper et al. (2007), "Some models and measures for evaluating performances with DEA: past accomplishments and future prospects," bellow we will describe how concepts embodied in two different definitions of efficiency and directed to very different types of problems are joined together by DEA to increase the power and scope of both. We then demonstrate how by using the original "fractional programming" DEA formulation, which embodies the engineering science definition, Charnes et al. (1978) extended the one-output-to-one-input ratio formulation to multiple outputs and multiple inputs by recourse to the concept of "Pareto efficiency" as used in welfare economics. We further describe how this fractional program gives way to an equivalent "linear programming" model with a dual model that effects the evaluations in a manner that embodies the *engineering science definition* in the "multiplier model" and the *Pareto (or Pareto-Koopmans)* definitions of efficiency in the dual or "envelopment model."

We shall recall that the DEA started with the mathematical interpretation of "engineering science" definitions of efficiency and used the duality theory of linear programming to show how, in DEA, they can be related to the Pareto-Koopmans definitions used in "welfare economics." Here, it would be helpful to remember that the basic ideas in DEA were first developed and applied to empirical data in the path breaking article by Farrell (1957). Intending to correct deficiencies found in productivity indexes, based on the "activity analysis" literature of Debreu (1951), and Koopmans (1951), he replaced the concept of productivity with the more general concept of "relative efficiency."

It is noteworthy that Farrell's uses of these ideas remained unnoticed, until after the publication of the seminal paper by Charnes et al. (1978) which extended and increased the scope: distinguishing between the "weakly efficient" performances, to which the Farrell-Debreu measure was limited, from the "fully efficient" or "strongly efficient" characterizations that they proposed.

By introducing a "fractional programming" formulation in which the ratios were restricted to lie between the values of zero and unity, Charnes et al. (1978) extended the usual engineering science single output-to-single input ratio measure of efficiency $\frac{Outputs}{Inputs}$ to multiple inputs and outputs without requiring recourse to a priori prescribed weights, such as are customarily used in engineering to deal with such cases.

DEA originated from the fractional programming formulation which (Model 9.3) generalizes the one-output-to-one-input ratio of engineering and science and (Model 9.2) derives its efficiency evaluations for each DMU relative to the performances of all DMUs.

$$
\begin{aligned}
&\max\ h_o(u,v) = \sum_r u_r y_{ro} \Big/ \sum_i v_i x_{io} \\
&\text{subject to} \\
&\sum_r u_r y_{rj} \Big/ \sum_i v_i x_{ij} \le 1 \text{ for } j = 1, \ldots, n, \\
&\text{with} \\
&\frac{u_r}{\sum_{i=1}^{m} v_i x_{io}}, \frac{v_i}{\sum_{i=1}^{m} v_i x_{io}} \ge \varepsilon > 0,
\end{aligned}
\tag{9.3}
$$

where y_{rj} and x_{ij} are observed values of outputs and inputs units relatively, $r = 1, \ldots, s$ and $i = 1, \ldots, m$ for each $j = 1, \ldots, n$ DMUs and the y_{ro} and x_{io} in the objective function represent the outputs and inputs for the DMU to be evaluated. A condition ε is a non-Archimedean element smaller than any positive real number is introduced to guarantee that solutions will be positive in these variables and each of them will be accorded "some" worth, however small.

Applying the prominent Charnes and Cooper (1962) transformation of variables that had led to the area of research that is now known as "fractional programming," see Schaible (1996), they converted this nonlinear and nonconvex problem to an equivalent linear programming problem (Model 9.5a) that is dual to the problem formulated by Farrell.

$$
\begin{aligned}
\mu_r &= \frac{u_r}{\sum_{i=1}^{m} v_i x_{io}} \ge \varepsilon > 0, \quad r = 1, \ldots, s. \\
v_i &= \frac{v_i}{\sum_{i=1}^{m} v_i x_{io}} \ge \varepsilon > 0, \quad i = 1, \ldots, m.
\end{aligned}
\tag{9.4}
$$

Using these variables presented in (Model 9.4), the problem in (Model 9.3) is transformed into the linear programming problem. This renowned transformation, in turn, provided *a previously unknown* contact with the efficiency concepts of welfare economics and of engineering and science. This in turn provides contact with the concept of "Pareto (1909) optimality," or "efficiency," as used in "welfare economics," and was later extended for use in "production economics" by Koopmans (1951) so that it is now referred to as Pareto-Koopmans efficiency:

Definition 1: Pareto—Koopmans efficiency is achieved by a DMU if and only if it is not possible to improve any of its inputs or output.

The most important property of this definition lies in the fact that (Model 9.3) it is not necessary to assign relative weights, etc., to determine the relative importance of the different inputs and outputs and (Model 9.4); its fulfillment is a necessary condition for optimality in any system where the weights that might be assigned are all positive.

Banker et al. (1984) showed that the CCR efficiency measure can be regarded as the product of a technical efficiency (BCC efficiency) measure and a scale efficiency measure matching with the economic notion of returns to scale. To estimate technical efficiency and the returns to scale, by addition of a convexity constraint BCC provided a modification of the CCR linear programming formulation. Banker et al. (1984) introduced the notion of most productive scale size (mpss), and showed how the CCR linear programming formulation can be employed to estimate the returns to scale.

CCR/CRS

Multiplier

$$\max z = \sum_{r=1}^{s} \mu_r y_{ro}$$

subject to

$$\sum_{r=1}^{s} \mu_r y_{rj} - \sum_{i=1}^{m} \nu_i x_{ij} \leq 0, \quad j = 1, \ldots, n; \tag{9.5a}$$

$$\sum_{i=1}^{m} \nu_i x_{io} = 1$$

$$\mu_r, \nu_i \geq \varepsilon > 0,$$

BCC/VRS

Multiplier

$$\max z = \sum_{r=1}^{s} u_r y_{ro} - u_0$$

subject to

$$\sum_{r=1}^{s} u_r y_{rj} - \sum_{i=1}^{m} v_i x_{ij} - u_0 \leq 0, \quad j = 1, \ldots, n;$$

$$\sum_{i=1}^{m} v_i x_{io} = 1 \tag{9.5b}$$

$$u_r, v_i \geq \varepsilon > 0,$$

$$u_o \quad \text{free in sign}$$

As highlighted by Cooper et al., (2007), in "Some models and measures for evaluating performances with DEA: past accomplishments and future prospects," the power and the computational convenience of DEA relies in the fact that the computations were executed not by the ratio form but also by the linear programming equivalent derived from the Charnes-Cooper transformation of fractional programming:

> the evaluations were made by means of the Pareto-Koopmans definition of efficiency, as in welfare economics, but were interpreted in terms of their engineering-science equivalents represented by the corresponding fractional form. **Thus both definitions of efficiency were brought into play and, as should be evident, this two-way contact can also make it possible for the engineering-science definition of efficiency to be applied to propositions in welfare economics, if desired**.

More specifically, because it generally has fewer constraints (Model 9.5a), it is easier to compute and since most computer codes use the simplex algorithm of linear programming which automatically generates solutions to both problems. Consequently access to the optimal μ_r and ν_i as well as the optimal λ_j and θ is thus provided without extra effort.

Both BCC/VRS and CCR/CRS DEA models represent convex production technologies. The input-oriented envelopment model takes the form:

CCR/CRS

	$\theta^* = \min \theta - \varepsilon\left(\sum_{i=1}^{m} s_i^- + \sum_{r=1}^{s} s_r^+\right)$	
	subject to	
Envelopment	$\sum_{j=1}^{n} x_{ij}\lambda_j + s_i^- = \theta x_{io} \quad i = 1, 2, ..., m;$	(Eq. 9.6a)
	$\sum_{j=1}^{n} y_{rj}\lambda_j - s_r^+ = y_{ro} \quad r = 1, 2, ..., s;$	
	$\lambda, \ s_i^-, \ s_r^+ \geq 0 \ \ \forall i, j, r.$	

BCC/VRS

	$\theta^* = \min \theta - \varepsilon\left(\sum_{i=1}^{m} s_i^- + \sum_{r=1}^{s} s_r^+\right)$	
	subject to	
Envelopment	$\sum_{j=1}^{n} x_{ij}\lambda_j + s_i^- = \theta x_{io} \quad i = 1, 2, ..., m;$	(9.6b)
	$\sum_{j=1}^{n} y_{rj}\lambda_j - s_r^+ = y_{ro} \quad r = 1, 2, ..., s;$	
	$\lambda, \ s_i^-, \ s_r^+ \geq 0 \ \ \forall i, j, r.$	
	$1 = \sum_{j=1}^{n} \lambda j$	(9.7)

where y_{rj} and x_{ij} are observed values of outputs and inputs of $r = 1, ..., s$ units relatively, and $i = 1, ..., m$ for each $j = 1, ..., n$ P/DMUs and the y_{ro} and x_{io} in the objective function represent the outputs and inputs for the DMU to be evaluated. Here, ε is a non-Archimedean element smaller than any positive real number, s_i^- and s_r^+ are slack values, θ^*.

The latter member of this dual pair of linear programming problems—i.e. (Model 9.6a)—is referred to as the "envelopment model." Applied successively to each member of a set of n DMUs it generates an "efficiency frontier" that "envelops" a "production possibility set." The first member of this dual pair—i.e., (Model 9.5)—is referred to as the "multiplier model" because it yields the dual variables that evaluate the outputs and inputs represented by the x_{io} and y_{ro} in a manner that differs from the ordinary uses of weights selected on an *a priori* basis.

As we can see from (Model 9.6a) and (Model 9.6b), the CCR model is the same as the "envelopment form" of the BCC model, except for the fact that it contains the condition $\sum_{j=1}^{n} \lambda j = 1$ is omitted. The $1 = \sum_{j=1}^{n} \lambda j$ (Model 9.7) constraints introduces an additional variable, μ_r, into the dual-multiplier form of the BCC model (Model 9.5b) obtained from the same data, which are then used in the following form:

In consequence, the variable u_0^*, which appears in the "multiplier form for the BCC model in (Model 9.4b) is omitted from the dual form of this CCR model (Model 9.4a)."

As noted earlier, the non-Archimedean $\varepsilon > 0$ is not a real number and hence cannot be assigned a numerical value. See Ali and Seiford (1990). However, it is not necessary to assign it a numerical value. Most DEA computer codes deal with Model 9.8a in a two-stage manner as follows: In stage one, the following modification of (Model 9.8a) is solved:

Farrell model

$$
\begin{aligned}
&\theta^* = \min \theta \\
&\text{subject to} \\
&\sum_{j=1}^{n} x_{ij}\lambda_j + s_i^- = \theta x_{io} \quad i = 1, 2, \ldots, m; \\
&\sum_{j=1}^{n} y_{rj}\lambda_j - s_r^+ = y_{ro} \quad r = 1, 2, \ldots, s; \\
&\lambda, \ s_i^-, \ s_r^+ \geq 0 \ \ \forall i, j, r.
\end{aligned}
\tag{9.8a}
$$

Envelopment model

$$
\begin{aligned}
&\max \sum_{i=1}^{m} s_i^- + \sum_{r=1}^{s} s_r^+ \\
&\text{subject to} \\
&\quad \sum_{j=1}^{n} \lambda_j x_{ij} + s_i^- = \theta^* x_{io} \quad i = 1, 2, \ldots, m; \\
&\quad \sum_{j=1}^{n} \lambda_j y_{rj} - s_r^+ = y_{ro} \quad r = 1, 2, \ldots, s; \\
&\lambda_j, s_i^-, s_r^+ \geq 0 \ \forall i, j, r,
\end{aligned}
\tag{9.8b}
$$

Thus, a solution with $\theta = \theta^*$ is obtained in which the slacks are omitted from the objective. Although this corresponds to Farrell's original formula tion, it does not solve (Model 9.4). To accomplish this, the following second-stage problem also needs to be solved, where the choices of s_i^- and s_r^+ cannot affect the optimal θ^* which was determined from (Model 9.8b). In this way, the optimal θ^* as determined in stage one is accorded "preemptive priority." This θ^* value cannot be reduced in exchange for increments in the slacks in (Model 9.8a), no matter how large are the sums that can be obtained with these slacks in the objective of (Model 9.8a). Hence, the non-Archimedean characterization of $\varepsilon > 0$ in the objective of (Model 9.8a) is preserved. Further, the maximizing objective used for the slacks in (Model 9.6) ensures that no alternate optimum for which $\theta = \theta^*$ can be chosen with all slacks zero when other alternate optima have positive slacks.

We shall specify here that model (Model 9.8a) is the one used in Farrell (1957), and therefore, it is occasionally referred to as the "Farrell model." In the economics portion of the DEA literature, it is often said to conform to the assumption of "free disposal" because it assigns a zero coefficient to every slack variable as part of the measure of efficiency and therefore omits these slacks from the objective, as in (Model 9.8b). In the operations research portion of the DEA literature, the "Farrell measure" is said to provide a measure of "weak efficiency" in order to reflect the fact that inefficiencies represented by the nonzero slacks are omitted from this measure.

It is noteworthy that both (Model 9.3a) and (Model 9.3) incorporate the engineering science definition of efficiency with the latter evaluating the denominator at unity. Incorporation of the Pareto-Koopmans definition (see below) of efficiency occurs in (Model 9.8a). *It is by the duality theory of linear programming, inherent in DEA models, that the two definitions are related to each other.*

The mathematical formulation of *Pareto—Koopmans efficiency* definition (1) takes the following formulation:

Definition 2: "Full" or "strong" (Pareto-Koopmans) efficiency is attained if and only if an optimal solution to (4) fulfills both of the following conditions:

(i) $\theta^* = 1$

(ii) All slacks are zero

"Weak efficiency" is attained if only condition (i) is to be satisfied as in a solution to (Model 9.6a).

Rajiv Banker, William Cooper, Lawrence Seiford, and Joe Zhu addressed the matter on returns to scale in DEA in Chapter Two in the Handbook on *DEA* (Kluwer Academic Publishers) for use for returns to scale and formulated important theorems presenting the interconnectedness of these two models.

Of particular relevance to our own analysis is the following theorem formulated by Banker and Thrall (1992) that submits that we can use the same projection to move all points onto the efficient frontier for (Model 9.6a) and proceed directly to return to scale characteristics for (Model 9.6a).

Theorem 1: The following conditions identify the situation for RTS for the CCR model given in (Model 9.6a):

(i) Constant return to scale prevail at $(\widehat{x_0}, \widehat{y_0})$ if $\sum_{j=1}^{n} \lambda_j^* = 1$ in any alternate optimum

(ii) Decreasing return to scale prevail at $(\widehat{x_0}, \widehat{y_0})$ if $\sum_{j=1}^{n} \lambda_j^* > 1$ for all alternate optima

(iii) Increasing return to scale prevail at $(\widehat{x_0}, \widehat{y_0})$ if $\sum_{j=1}^{n} \lambda_j^* < 1$ for all alternate optima.

Theorem 2: Sign condition for BCC and CCR models:

(i) The case of constant return to scale. $u_0^* = 0$ for some solutions to (Model 9.5b) if and only if $\left(\sum_{j=1}^{n} \lambda_j^* - 1\right) = 0$ for some optimal solutions to (Model 9.6a)

(ii) The case of decreasing returns to scale. $u_0^* > 0$ for all optimal solutions to (Model 9.5b) if and only if $\left(\sum_{j=1}^{n} \lambda_j^* - 1\right) > 0$ for all optimal solutions to (Model 9.6a)

(iii) The case of increasing return to scale. $u_0^* < 0$ for all optimal solutions to (Model 9.5b) if and only if $\left(\sum_{j=1}^{n} \lambda_j^* - 1\right) < 0$ for all optimal solutions to (Model 9.6a)

The later theorem removes the possibility that uses of the CCR and BCC models might lead to different RTS characteristics. It is also remarkable because difference might be expected from the fact that (Model 9.5) effects its evaluation locally with respect to a neighboring facet while (Model 9.3b) *effects its evaluation globally with respect to a facet (or point) representing MPSS* (see the Handbook on Data Envelopment Analysis by Cooper, Seiford Tone (Kluwer Academic Publishers, p. 57).

If certain DMUs were efficient with VRS and not with CRS, this means that the use of CRS has added new information. They were technically efficient, as shown by VRS, but not scale efficient when CRS and VRS are used jointly in this manner.

9.6 Measuring sustainable development goals performance in the age of globalization

Herbert Simon was awarded a Nobel Prize in 1978, for "his pioneering research into the decision-making process within economic organizations" and had focused on decision-making and problem-solving processes, using computers to simulate human thinking. In his contribution to "The State of Economic Science: Views of Six Nobel Laureates," Simon identified two assumptions about utility maximation widely accepted by nearly all economists. Every "person," or we would call here, a decision or policy maker ought to maximize utility, as uniquely rational behavior:

(1) *possesses a consistent, comprehensive utility function that orders all the possible alternatives of choice, and*

(2) *selects among alternatives so as to maximize subjective expected utility.*

Within this rationality context, an exchange between persons is permitted, or acceptable, only if both agree on a "price" whereby their individual utility will be increased by the exchange, and so everyone will be better off. As Simon puts it: "when certain other conditions are also satisfied, the exchange is said to produce a Pareto optimum, a situation that could not be altered in any way to produce an increase in utility for both participants. Economists also refer to solutions of decision problems that are Pareto optimal as producing an efficient allocation of resources."

Referring the "Das Kapital" of Marx, the "General Theory" of Keynes, the work of rational expectationists like Lucas and Sargent, and the monetarists, like Friedman, and his fellow co-authors, Simon places "utility maximization" at the heart of economic thought:

> *the allegiance of economists of all sects and descriptions to subjective expected utility maximization is hardly less unanimous than the allegiance of physicists to Newton's Three Laws of Motion in their relativised form.*

He expands his coverage to utility function, specifically highlighting the gender dimension, particularly addressing the "utility function of women" that, according to him, "has shifted, for" over time "as a consequence of consciousness raising, so that they prefer outside work to homework more than they did before." Reverting to the scientific achievements during the recent decades, in the domain of cognitive psychology, Simon acknowledged the emergence of a large body of empirically tested theory of decision-making and problem-solving that substitutes *satisficing* for *optimizing*. This shift in thinking gives central position *to the limits of human ability to deal with real-world complexity*, and to the methods of radical approximation that are used to remove this mismatch *of adaptive power to problem difficulty:*

> enlarging the scope of economic decision theory, it does not assume that the alternatives of choice are given a priori, but shows how they can be generated by selective (heuristic) search in problem spaces. It is beginning to investigate the processes that focus human attention on certain problems so that others are ignored, and the processes of representing or framing.

Impressed by the work of cognitive psychologists in studying the decision-making and problem-solving abilities of humans to make rational choices, Simon and Arrow both contend that economists should pay closer attention to the achievements of their work.

I could not agree with them, more in that, these new theoretical and methodological developments in psychology, suggesting to use computers to simulate complex human thought processes and thereby to model problem-solving and decision-making situations of real-world complexity, have now created "a firm foundation on which economic research can proceed."

In his Nobel Memorial Lecture, delivered on December 8, 1978, entitled Rational Decision-Making in Business Organizations, Simon concluded:

> We may be able to reach consensus that a certain criterion of rationality is appropriate to a particular game, but if someone challenges the consensus, preferring a different criterion, we will have no logical basis for persuading him that he is wrong.
>
> ***(Simon, 1978)***

With the present book, we have entered into uncharted water, and what we could analyze and present in the current volume so far, is only a drop in the ocean of possible DEA-based applications toward evaluating the performance of SDGs achievements in diverse contexts and levels. Measuring the performances of countries and other DMUs, in the age of globalization, is a very difficult task, not only because of difficulties associated with the choices and definitions of the performance variables but also given the new circumstance, in trying to reach multiple development objectives simultaneously under the pressure of post-COVID crisis rehabilitation.

To deliver on "Triple Bottom line, commonly known as, 3Ps: People, Planet, and Profit," the fundamental questions of, what I would call 3Es: *Efficiency*, *Effectiveness*, *Equity*, assessments will continue to occupy the international development and research agenda for the years and decades to come, until we arrive to the Future We Want and reach our sacred dream: a sustainable globalized world free of poverty, with peaceful, prosperous, and happy societies:

- *Efficiency—How successful are countries and other policy and DMUs in achieving* "more sustainability benefits from globalization for less cost"*?*
- *Effectiveness—What is distance between observed impact of their intervention and a set of desired development goals?*
- *Equity—How fair are they in allocating the resources or distribution of benefits?*

As demonstrated in our empirical applications, answering these questions calls for a shift in analytical thinking from economic efficiency toward socioeconomic effectiveness and sustainability impact analysis. Considering the large number of SDGs and development targets, the accomplishments and failures of various PMUs and DMUs and their policy performance will always be multidimensional. Once we have agreed on a single measure of such progress, it becomes possible to document numerically how those policy and decision makers succeed (or fail) in achieving their goals and communicate them politically.

The United Nations 2030 Agenda for Sustainable Development by establishing the 17 Sustainable Development Goals formulated an ambitious global framework for tackling the challenges our world faces today. The Agenda addresses the interconnected goals of achieving economic prosperity, social justice, and environmental sustainability for all people, with a pledge to leave no one behind.

Already at the time of elaboration of Agenda 2030, the UN Department of Economic and Social Affairs, Division for Sustainable Development in their *Prototype Global Sustainable Development Report* (United Nations, 2014) recommended that:

> Indicators corresponding to the future SDGs are most important for monitoring future progress, but they will need to be complemented by composite indices of sustainable development progress.
>
> Sustainable development indicators derived from a set of agreed international goals or commitments, and a composite indicator, which is the compilation of

individual indicators into a single index, are considered to be a good vehicle in helping to measure and monitor sustainable development and progress achieved towards it.

All these indicators are meant to present complex data and trends in simplified form to policymakers. They can inform policy formulation on the basis of information that is transparent and evidence-based. The challenges, among others, are to develop and agree upon the fully integrated framework of measurement at the global level, which includes both goals and a set of indicators for - and assessing the needs and tracking the progress of - sustainable development.

The SDGs as they stand, entail a task to improve the system of indicators for progress measurement and under *Goal 17—Strengthen the means of implementation and revitalize the global partnership for sustainable development*, as a distinct target 17.19 explicitly foresee:

17.19 by 2030, build on existing initiatives to develop measurements of progress on sustainable development that complement GDP, and support statistical capacity building in developing countries

There is a universal unanimity that achieving these global goals requires specific targets and measurable indicators, informed by timely and reliable data, that will enable policy-makers to analyze the progress and make informed decisions about the best path forward and so their achievements. The availability of statistical data has always been at the center of the UN Statistical Commission deliberations.

As a mean of addressing the methodological issues and standards, in 2015, the Commission adopted a three-tiered framework of 232 unique global indicators proposed by its Inter-Agency Expert Group on SDG Indicators (IAEG-SDGs). All development stakeholders identified this as an important step forward for generating internationally comparable harmonized data that will considerably facilitate the implementation of the global indicator framework.

IAEG-SDGs classified all indicators into three tiers based on their level of methodological development and the availability of data at the global level, as follows:

Tier Classification Criteria/Definitions:

Tier 1: Indicator is conceptually clear, has an internationally established methodology and standards are available, and data are regularly produced by countries for at least 50% of countries and of the population in every region where the indicator is relevant.

Tier Classification Criteria/Definitions: (*cont'd*)

Tier 2: Indicator is conceptually clear, has an internationally established methodology and standards are available, but data are not regularly produced by countries.

Tier 3: No internationally established methodology or standards are yet available for the indicator, but methodology/standards are being (or will be) developed or tested.

Source: UN Statistics Division

The IAEG-SDGs acknowledged that all indicators are equally important, and the establishment of the tier system is intended solely to assist in the development of global implementation strategies. They recognize that for tier I and II indicators, the availability of data at the national level may not necessarily align with the global tier classification and countries can create their own tier classification for implementation.

Thanks to the positive developments as of the 51st session of the UN Statistical Commission, held on 3–6 March 2020, the global indicator framework does not contain any Tier III indicators. As of December 28, 2020, based on the data availability review at the 11th meeting, the updated tier classification contains 130 Tier I indicators, 97 Tier II indicators, and four indicators that have multiple tiers (different components of the indicator are classified into different tiers).

To better address local circumstances, these global indicators are complemented by national and regional indicators, which are driven by national and regional priorities. To ensure the highest level of ownership, national statistical systems are assigned the central responsibility of compiling data and metadata according to the fundamental principles of official statistics (relevance, impartiality, equal access, accountability, transparency, confidentiality, etc.). In the framework of the global information database, local data is being made available for further reporting initiatives at the national, regional, and global levels.

Within a well-established collaborative system, regional mechanisms facilitate the transmission of data and metadata from the national to the global level, while international agencies provide internationally comparable data in different statistical domains, calculate global and regional aggregates, and provide accompanying metadata for global reporting. Countries are offered an opportunity to present their data to the international community through their Voluntary National Reviews, annually, each July at the United Nations High-level Political Forum (HLPF). This review mechanism enables them to present their efforts and specific actions employed to achieve the goals, and the identified challenges hindering the implementation they have encountered.

With a remarkable progress in methodological developments and data collection and sharing, even with this substantial reporting system, for ensuring a successful SDG monitoring and evaluation, there remain important challenges to be addressed:

Methodological: A number of official indicators lack agreed-upon methodologies or available official data.

- *Local capacity:* Some countries' national statistical offices lack the capacity to produce relevant data.
- *Legitimacy:* While private tech companies produce and have a great potential for providing big data that opens up huge opportunities for measuring crucial SDG indicators, the questions around unofficial sources of these data, and their regularization for the use in the formal reporting system.
- *Coverage*: The official indicator framework don't cover—how, for instance, do we measure the "spillover" effect of one country's SDG action on another country's progress?

To be able to fully understand and evaluate the progress on the SDGs, the research and development community needs as much reliable data as possible, ensuring that everyone is counted and no one is left behind. There is a strong agreement that policy-makers and stakeholders at all levels need access to the real time data in a user-friendly way, so that they have the best available information to make meaningful—and measurable—progress on the SDGs.

The Agenda 2030 general vision—"achieving sustainable development in its three dimensions—economic, social, and environmental"—describes not only the development process but, firstly, the result achieved. Multiple dimensionality principle was already defined in the first Human Development Report published in 1990 (UNDP, 1990): entitled "Concept and Measurement of Human Development," with the acknowledgment of the necessity to consider all indicators in the sustainable human development measurement process:

> Human development has two sides: the formation of human capabilities—such as improved health, knowledge and skills—and the use people make of their acquired capabilities.

The OCED/JRC *Handbook on Constructing Composite Indicators* (OECD, 2008) sets grounds for defining and calculating composite indicators and provides some key aspects of their uses.

- *In general terms, an indicator is a quantitative or a qualitative measure derived from a series of observed facts that can reveal relative positions* (e.g., *of a country) in a given area.*
- *When evaluated at regular intervals, an indicator can point out the direction of change across different units and through time. In the context of policy analysis, indicators are useful in identifying trends and drawing attention to particular issues.*
- *They can also be helpful in setting policy priorities and in benchmarking or monitoring performance.*

The Handbook also presents a checklist for composing composite indicators, encouraging the involvement of experts and stakeholders during the Steps 1 and 2 of the process (Table 9.1).

> A composite indicator is formed when individual indicators are compiled into a single index on the basis of an underlying model. The composite indicator should ideally measure multidimensional concepts which cannot be captured by a single indicator, e.g. competitiveness, industrialisation, sustainability, single market integration, knowledge-based society, etc.

Table 9.1 Checklist for building a composite indicator.

1. Theoretical framework
   - To get a clear understanding and definition of the multidimensional phenomenon to be measured.
   - To structure the various subgroups of the phenomenon (if needed).
   - To compile a list of selection criteria for the underlying variables, e.g., input, output, process.
2. Data selection
   - To check the quality of the available indicators.
   - To discuss the strengths and weaknesses of each selected indicator.
   - To create a summary table on data characteristics, e.g., availability (across country, time), source, type (hard, soft or input, output, process).
3. Imputation of missing data
   - To estimate missing values.
   - To provide a measure of the reliability of each imputed value, so as to assess the impact of the imputation on the composite indicator results.
   - To discuss the presence of outliers in the dataset.
4. Multivariate analysis
   - To check the underlying structure of the data along the two main dimensions, namely individual indicators and countries (by means of suitable multivariate methods, e.g., principal components analysis, cluster analysis).
   - To identify groups of indicators or groups of countries that are statistically "similar" and provide an interpretation of the results.
   - To compare the statistically determined structure of the data set to the theoretical framework and discuss possible differences.

(*Continued*)

Table 9.1 Checklist for building a composite indicator.—cont'd

5. Normalization
   - To select suitable normalization procedure(s) that respect both the theoretical framework and the data properties.
   - To discuss the presence of outliers in the dataset as they may become unintended benchmarks.
   - To make scale adjustments, if necessary.
   - To transform highly skewed indicators, if necessary.
6. Weighting and aggregation
   - To select appropriate weighting and aggregation procedure(s) that respect both the theoretical framework and the data properties.
   - To discuss whether correlation issues among indicators should be accounted for.
   - To discuss whether compensability among indicators should be allowed.
7. Uncertainty and sensitivity analysis
   - To consider a multimodelling approach to build the composite indicator, and if available, alternative conceptual scenarios for the selection of the underlying indicators.
   - To identify all possible sources of uncertainty in the development of the composite indicator and accompany the composite scores and ranks with uncertainty bounds.
   - To conduct sensitivity analysis of the inference (assumptions) and determine what sources of uncertainty are more influential in the scores and/or ranks.
8. Back to the data
   - To profile country performance at the indicator level so as to reveal what is driving the composite indicator results.
   - To check for correlation and causality (if possible).
   - To identify if the composite indicator results are overly dominated by few indicators and to explain the relative importance of the subcomponents of the composite indicator.
   - To conduct sensitivity analysis of the inference (assumptions) and determine what sources of uncertainty are more influential in the scores and/or ranks.
9. Links to other indicators
   - To correlate the composite indicator with other relevant measures, taking into consideration the results of sensitivity analysis.
   - To develop data-driven narratives based on the results.
10. Visualization of the results
    - To identify a coherent set of presentational tools for the targeted audience.
    - To select the visualization technique which communicates the most information.
    - To present the composite indicator results in a clear and accurate manner.

Launched in 2015 by the Sustainable Development Solutions Network (SDSN) and the Bertelsmann Stiftung, the annual Sustainable Development Report is the first worldwide study to assess where the world stands with regard to achieving sustainable development. The report contains the SDG Index and Dashboards, which use a combination of official and unofficial data and indicators to understand each country's progress on the 17 SDGs. Each year, a report is published on the degree of compliance with each of the SDGs, using a set of selected indicators, and an overall index for each country, known as the SDG Index, which is scored between 0 and 100 according to a weighting of the indicators for each objective (SDSN TReNDS, 2020) (Table 9.2).

Table 9.2 Recent composite indicators constructed by Sustainable Development Solutions Network and others.

Created by SDSN	Created by others
1. 2020 Sustainable Development Report	**1.** SDG India Index & Dashboard. NITI Aayog.
2. World Happiness Report 2020	**2.** Transforming Australia: SDG Progress Report. SDG Transforming Australia Project.
3. 2020 Europe Sustainable Development Report	**3.** OECD Better Life Index (Interface available in English, Spanish, French, Russian, German, Portuguese, and Italian.)
4. 2020 Africa SDG Index and Dashboards Report	**4.** 2020 Social Progress Index. Social Progress Imperative.
5. Atlas municipal de los objectivos de desarrollo sostenible en Bolivia 2020	**5.** 2019 Corruption Perceptions Index. Transparency International. (Interface available in English, French, Spanish, Russian, and Arabic.)
6. 2020 SDG Cities Index for Italy	**6.** Human Development Report 2019 and Human Development Index Ranking. UNDP.
7. 2019 SDG Index and Dashboards Report for European Cities	**7.** 2019 Global Multidimensional Poverty Index. UNDP.
8. 2019 Arab Region SDG Index and Dashboards Report	**8.** 2020 Gender Social Norms Index. UNDP.
9. 2019 Africa SDG Index and Dashboards Report	**9.** 2019 Corporate Tax Haven Index. Tax Justice Network.
10. 2019 US Cities Sustainable Development Report	**10.** 2020 Financial Secrecy Index. Tax Justice Network.
11. 2019 SDG Index for Latin America and the Caribbean	
12. Informe: Los ODS en 100 ciudades españolas	
13. 2019 Sustainable Development Report—Mediterranean Countries Edition	
14. 2018 Sustainable Development Report of the United States	

Multiple international development institutions and nongovernmental initiatives had been undertaken after the adoption of SDG toward the construction of composite indicators (Fig. 9.1). With about 10 mln annual revenues, the UN SDSN is the leading network in developing and disseminating the SDG Indexes with diverse coverage: global, regional, country, etc.

This second annual report from the SDSN including US city-level data from 2018 Index and Dashboards ranks the 100 most populous metropolitan areas in the United States on the SDGs. It is particularly relevant for urban leaders whose engagement is crucial for sustainable development, as 85% of the US population live in cities. In 2018 Espey et al. initiated Cities SDGs Index. in Leaving No US City Behind: The 2018 US Cities SDGs Index. and in 2019 Lynch et al. published The 2019 US Cities Sustainable Development Report.

A large body of literature also exists examining the environmental impacts of the household behavior. Poom and Ahas, in "How Does the Environmental Load of Household Consumption Depend on Residential Location?" present a number of studies. We chose few of those that contain some interesting conclusions. The environmental impacts of the household expenditure structure in the EU were studied by Liobikiene and

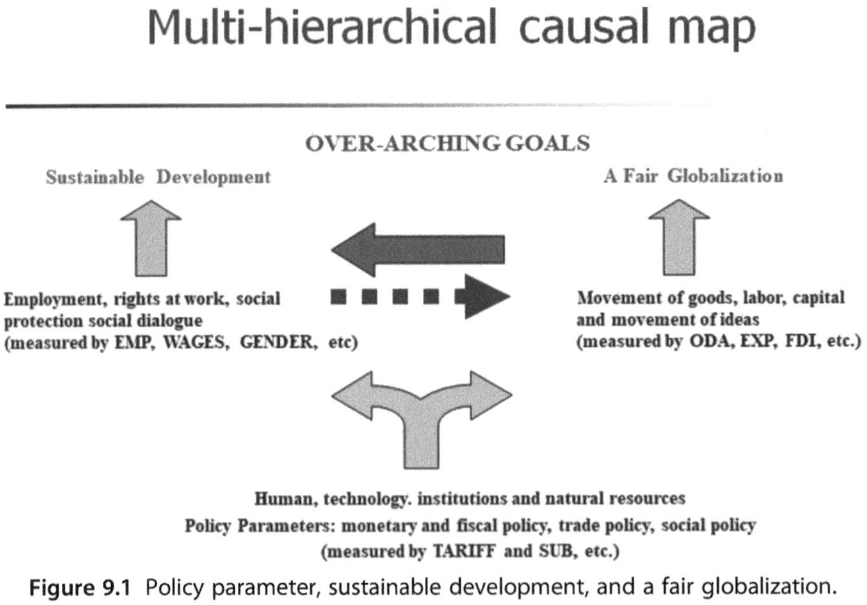

Figure 9.1 Policy parameter, sustainable development, and a fair globalization.

Mandravickait (2013). (1) food and beverages, transport, and housing were found to be the most polluting ones in terms of greenhouse gas emission and acidifying compound emission intensities; (2) clothing and footwear, furnishings and household equipment, and restaurants and hotels were found to be moderately polluting; (3) lowest intensities of greenhouse gas emissions were attributed to health, communications, education, recreation and culture, and miscellaneous good and services. The authors concluded that it is not possible to achieve more sustainable consumption patterns by only changing the household consumption structure. Governmental policies affecting technological development and inducing more sustainable production and consumption are required as well.

Mobility, shelter, and food were identified to be the most influencing categories for environmental footprints in terms of greenhouse gas emissions, land use, water use, and material use by Ivanova et al. (2015) based on the analysis of the environmental impact of household consumption using data from 43 countries. Shelter, transport, and food and nonalcoholic beverages were found to be the three main causes of the carbon load of household consumption in Estoniar by Poom and Ahas (2016). The choice of diet, the use of a private car, and the household size were found to have a significant influence on consumption-related environmental impacts in Denmark by Kalbar et al. (2018). Girod et al. (20140) reviewed the carbon footprints of products in five main categories including food, shelter, transportation, goods, and service. According to their findings, in all of the categories, there are consumption options, which enables the limiting the global temperature rise to 2°C.

Sustainability Composite Indexes are also under construction for individual SDGs—Kynčlová et al., 2020 created composite indexes at the level of individual SDGs. They introduced an index to measure the progress toward meeting the industry-related targets of the 2030 Agenda. Their proposed SDG-9 index benchmarks industrial performance while preserving the environment and promoting inclusiveness. The final ranking compares performance of 128 economies over the period 2000–2016. Economic growth should be accompanied by technological changes and efficient use of natural resources.

Prieto and Zofio (2001) used DEA to evaluate the effectiveness of Spanish municipalities in achieving goals (specified by the central government) in areas such as the provision of water quality, sewage treatment, paving and lighting and artistic and sporting facilities, thus expanding the DEA beyond "efficiency" evaluations to "effectiveness." See p. 66 in Cooper et al. (2000) for definitions of these two concepts and their relations to each other.

The Association for Supporting the SDGs for the United Nations (ASD), an international NGO that enjoys special consultative status with the ECOSOC, annually announces the SDGs Business Index andand Sustainable Business (SDGBI) that analyzes and evaluates the corporate sustainability of 3000 companies around the world. The SDGBI is an SDGs-based Corporate Sustainability Analysis Index for global corporations, submitted as a written statement (E/2019/NGO/91) at the High-Level Segment of the 2019 HLPF. Since 2016, the ASD has been announcing the SDGBI, and recently it has been introduced as a tool of sustainable business engagement and ESG evaluation index by UN ESCAP's "SDG Help Desk" "Sustainable Business Tools and Methodologies."

The OECD Action Plan, endorsed by the OECD Council in December 2016, underscores the organizations commitment to supporting the achievement of the SDGs describes how the OECD will support the 2030 Agenda through its legal instruments, its expertise in policy analysis and its know-how on statistics, indicators, and systems for monitoring performance. As part of this extensive plan, the OECD Statistics and Data Directorate developed a unique methodology for measuring the distance that OECD countries would need to travel in order to meet the SDG targets.

The study, *Measuring Distance to the SDG Targets 2019: An Assessment of Where OECD Countries Stand* is closely aligned with the *UN Global Indicator List for the SDGs*, and, wherever possible, reflects the level of ambition agreed by Member States when setting the 2030 Agenda. It was first released as a pilot study in 2016; following feedback from OECD member countries and work to expand indicator and country coverage, a revised and expanded version was published in June 2017. This 2019 edition of the Study further expands the indicator, target, and country coverage.

An earlier version of this Study was presented to the OECD Council on the 2030 Agenda for Sustainable Development in March 2019. Based on the UN Global List of 244 indicators, the study evaluates the distance that OECD countries need to travel to meet SDG targets for which data are currently available. The 2019 edition presents the latest results for OECD countries, both on *average* and individually, as well as new exploratory approaches to assessing progress over time and transboundary aspects embodied in the Agenda. By providing a high-level overview of countries' strengths and weaknesses in performance across the SDGs, this study aims to support member countries in navigating the SDGs and in setting their own priorities for action within the broad 2030 Agenda (Fig. 9.2).

Towards DEA composite indicator

'Best Practice'

Within Sample of Countries for the one that yields the Highest Possible TOTAL for the given set of weights

$$I_{io} = \frac{\sum_{r=1}^{s} w_r X_{io,r}}{\max\limits_{X_i \in \{studied\,countries\}} \sum_{r=1}^{s} w_r X_{i,r}}$$

'Benefit-of-the-doubt'

Weights are chosen so that the evaluated country has a MAXIMAL Composite Indicator Value

$$I_{io} = \max_{w_r} \frac{\sum_{r=1}^{s} w_r X_{io,r}}{\max\limits_{X_i \in \{studied\,countries\}} \sum_{r=1}^{s} w_r X_{i,r}}$$

Figure 9.2 Weights in best practice versus benefit-of-the-doubt approach.

DEA is a benchmarking tool for finding the best production/transformation practices. The ultimate goal is to improve these processes of inefficient DMUs/countries given their mix of resources and products in comparison with the countries on the frontier with a similar mix of indicators. The optimal DEA solutions that are derived from the mathematical program are based on the Benefit-of-the-doubt (BOD) versus a priori prefixed weights approach (see Fig. 9.2 above). BOD indicates the importance attached to individual input/output variables for which a country has a comparative advantage and performs relatively best compared to others in the sample.

In DEA models, a multi-Input multi-Output situation is reduced to a ratio of a single "virtual" output and a single "virtual" input by applying multipliers or weights to observed inputs and outputs. In DEA models, weights are chosen so as to maximize the efficiency of the PMU (country) under consideration. Weights are model solutions that maximize the index value for each country, subject to the constraint, that when those same weights are applied to the data of any other country, that other countries' efficiency ratio cannot exceed "unity."

That is, no PMU (country) is permitted to have an efficiency score higher than 100%. The endogenously selected weights maximize the weighted sum of the index value of the country performance and provide information that can be used to guide trade-offs between different inputs/outputs or to determine the results of increases/decreases in their corresponding amounts. Parting ways with conventional index calculations (see

Maragos and Despotis (2002)), however, rather than using fixed and predetermined weights throughout, we have determine the optimal weight to be given to each factor, thus enabling us to form the ratio between the optimally weighted outputs and the optimally weighted inputs. For both the outputs and the inputs, an index of the component factors has to be constructed. The present approach provides multiple advantages in terms of political acceptance and easy to communicate.

One interesting advise from William Cooper was to guide choice of "desirables" and "undesirables" using the ratio form $\frac{Outputs}{Inputs}$. Thus, "% of work force employed" would go into the numerator as a "desirable." The "% of child population in labor force" would go into the denominator as an "undesirable,"i.e., increases in the former increase the value of the ratio and increases in the latter would decrease the value of the ratio. Optimal value of the ratio form of multiple outputs (=desirables) and multiple inputs (=undesirables) transform into the usual DEA output and input-oriented models. Here, however, one will need to use it for conceptual guidance in their choice of inputs and outputs.

9.7 Agenda 2030 in post-COVID-19 global reset

Until December 31, 2019, when the WHO first learned about, what then was to be called, COVID-19, SARS-CoV-2 new virus, following the report of a cluster of case of "viral pneumonia" in Wuhan, China, the world was well underway on implementing of the Agenda 2030 adopted by the UN.

In our earlier studies, for our analysis, we have selected limited number of policy and performance variables, at our own discretion, for which the data were available at that point in time. Needless to say that with our prominent methodological developments, our approach must be seen as setting forth a proposition. While the analytical framework remains solidly adept to evaluate the double overarching goals of Sustainable Development and a Fair Globalization, any new scenario, that will include divergent set of input-output sets would be most welcome and will yield a different results.

As a black-box, nonparametric tool, a particular quality inherent in the DEA, the framework provides a unique opportunity for using the real time statistical data, that is progressively becoming available, on all the relevant domains presented in Fig. 9.1 ODA, exports, foreign investments, trade in goods, migration, etc., all the dimensions of the Decent Work and Social and Sustainable Development and enabling policy parameters, including

Figure 9.3 Policies for sustainable globalization in post-COVID-19 global reset.

institutional, human, technological, and natural resources. To perform county comparison in the context of globalization reform, the ultimate objective of the simulation would be to document past performance, establish baseline for gauging improvements, and make valid comparison across countries to (Fig. 9.3):

- *optimize policy operations to improve country performance*
- *quantify progress toward meeting SDGs*
- *explore the potential determinants of inefficiency*
- *identify high-performing countries whose practices might be adopted by others, expand networks for learning*
- *implement new approaches; create enabling regulations to fine-tune targets and incentives.*

Since December 2019, the entire world has mobilized its financial, intellectual, and physical capacity toward the understanding and the discovery of the mechanisms by which SARS-CoV-2 causative agent of the COVID-19 pandemic hijacks human cells and manipulates to favor the production of its proteins over the cell's own ones. Coronavirus has equally hijacked the development cooperation and international research agenda. The tiny virus could "immediately achieve" policy change, what thousands and millions engaged in antiglobalization movements and environmental militants were unable to realize during the long decades of their advocacy.

The catastrophe that was imposed on us, demonstrated how unprepared and fragile our governance systems are, both in developed and developing world alike. The unprecedented human-centered efforts of national governments that were (some still are) individually and collectively copying with this global tragedy, as a litmus paper, helps us to highlight and confirm our long standing belief, that *policies are made by people for people* and must adapt and introduce immediate remedying measures where necessary to review, adjust, and change policy priorities, to relocate, review budget priorities, to redistribute resources where needed to remedy any adverse impact of any intervention and create environment for meeting human needs.

With its appearance, COVID profoundly challenged the foundation of definition of the notion of globalization! Conventional definition of globalization would comprise four main domains: Movement of Goods, Movement of Labor, Movement of Capital, and Movement of Ideas, whereas, today, we have a new dominant dimension of globalization, which overshadowed everything else. Being both as a consequence—an "Output" of globalization and also a cause—an "Input" to further globalization, identification of a "right role" of COVID, for policy analysis, in terms of its impact on/from globalization will need significant expert discussions (see Fig. 9.3). COVID has clearly altered the development agenda and all ongoing and upcoming international deliberations toward the implementation of Agenda 2030 during the Global post-COVID Reset will need to center on interconnectedness of COVID and SDGs.

In our analysis, we have only used the data covering the monetary dimension of the poverty and used 1 dollar/day as an indicator for DEA analysis. But it is noteworthy that the multidimensional nature of poverty was first presented by Townsend (1979) already in 1979 and further developed by Chambers (1983). The multidimensional poverty definitions advocated by the DAC of the OECD in its guidelines for poverty reduction (OECD, 2001) have now been widely accepted. Most development cooperation agencies embody the DAC guidelines in their operational and research activities when fighting poverty.

The DAC recognizes that poverty is defined in various ways but suggests that poverty definitions should be context-specific and should encompass the key deprivations faced by poor women and men and identify the ways in which they are incapacitated (see Fig. 9.4). The DAC has adopted the capabilities approach to poverty, building on work by Sen (1985,

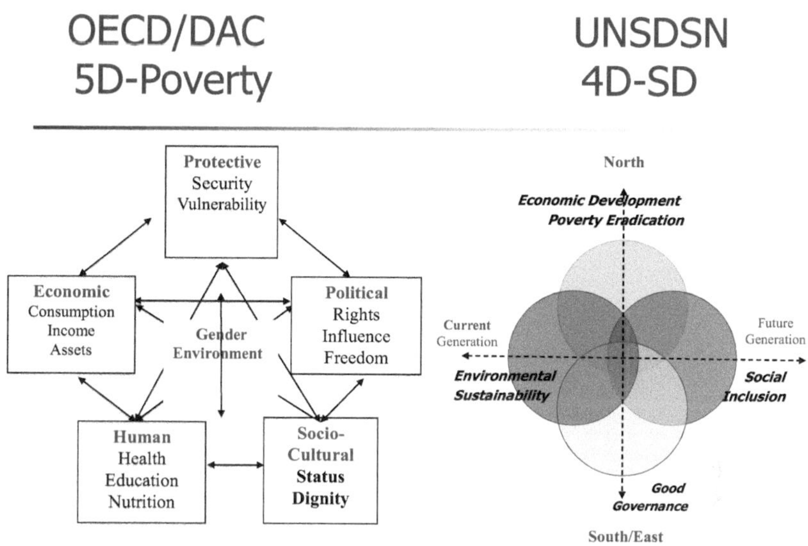

Figure 9.4 The multiple dimensions of poverty and sustainable developlement OECD (2001).

1999) and Martha Nussbaum (Nussbaum, 2000; Nussbaum and Sen, 1993) and suggests that poverty outcomes are affected by individual and household capabilities and that these capabilities are interrelated. The DAC emphasizes that distinguishing between capabilities is important both for analysis and in order to design effective policies and interventions.

The links indicated by arrows in Fig. 9.4 are significant. Each box represents an important dimension of poverty, which affects—and is affected by—all the others. Household members may consume little and be vulnerable partly because they lack assets, often because of inadequate income, poor health and education, or because they lose their few productive assets as a result of shocks. Lack of human rights and political freedoms indicates a risk of violent conflict shocks. Vulnerability and social exclusion hamper human and political capabilities, reducing incomes and assets, and so on. The fact that different dimensions of poverty are tightly interrelated, while still distinct and imperfectly correlated, is a major reason for a multidimensional concept.

Poverty, gender, and environment are mutually reinforcing, complementary and cross-cutting facets of sustainable development. Mainstreaming gender is key to reducing poverty across its various dimensions, and also to improving the environment. Environmental factors influence gender relations as well as poverty outcomes. Gender equality and environmental sustainability are international development goals in their own right.

In fact it is the added value of our model to the literature in this domain that we expanded the notion of efficiency of DMUs to the effectiveness of PMUs. The model transcends standard measures to more comprehensive economic, social, and environmental attributes.

- *Better description of reality than equal weighting: Weights are not pre-assigned, they are calculated from data.*
- *Sensitive to national policy priorities: Weights are endogenously determined by the observed performance.*
- *No other weighting scheme yields higher composite score (such that overall relative performance index is as high as possible).*
- *Benchmark is not based upon theoretical bounds, but a linear combination of observed best performances incentive generating rather than punishing laggards.*
- *Weights may help to define trade-offs between policy priorities.*

DEA was identified by the OECD/JRC. (2008) as a new approach to measuring and evaluating the many dimensions of economic and/or social performance. "It is designed to provide a summary measure of the actual performance of each entity (called a DMU) as evaluated relative to the performances of all of the other potentially pertinent DMUs and, simultaneously, to identify the sources and amounts of each of the deficiencies in the performance of every DMU."

"Efficiency Evaluation: Extends the classical "Pareto-Koopmans" definition of "efficiency" as used in "Welfare Economics" to a more general "Social-Economic Welfare" performance measure as in the following: Full (100%) efficiency is attained by a DMU if and only if none of its inputs or outputs can be improved without worsening some of its other inputs or outputs." (Note: this definition avoids the need for weights or other preassigned measures of relative value that might otherwise be needed to reflect relative merits that are assumed to apply to different inputs and outputs).

We made an attempt to propose a Comprehensive Conceptual Framework (Fig. 9.5) that can be used by Global Decision Makers (UNDESA/DPADM) in order to foster integrated approaches to the implementation of the 2030 Agenda in the context of globalization. It will enable an analyst to map the existing SDG indicator into corresponding unit: individual, household, community, city, region, or global; and put into correspondence with appropriate level op Sustainable Development Strategic Papers and Plan. Once specific set of policy variables and performance indicators are finalized by the respective stakeholders' and experts group, an appropriate DEA Model can be applied to draw solutions that will identify inefficiencies and will guide future improvements. Gender aspects, generation dimension, and time prospects can also be addressed in the framework of an additional simulations (Fig. 9.5).

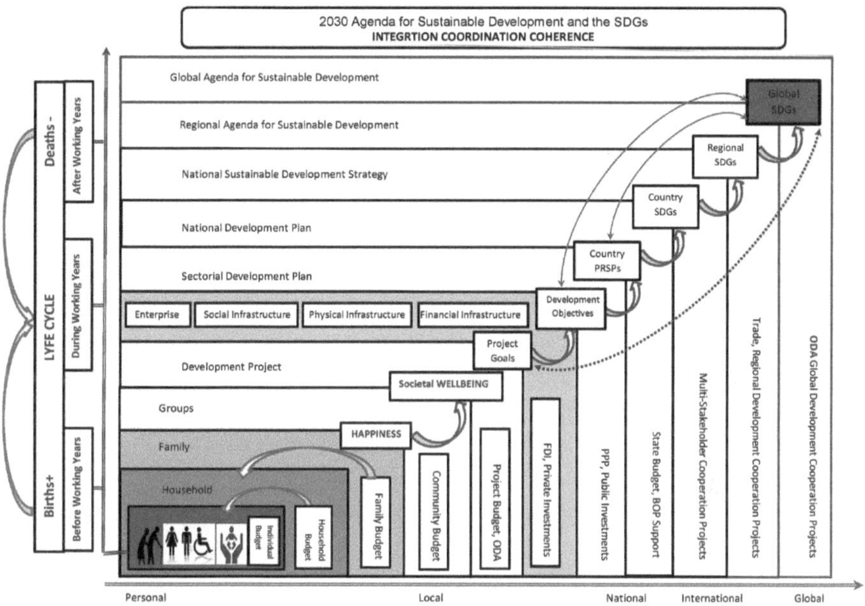

Figure 9.5 Integration, coordination, and coherence for agenda 2030 and sustainable development goals.

9.8 The Geneva Consensus Foundation: a tribute to William W. Cooper

Our earlier book "Diagnostics for a Globalized World," was dedicated to the memory of William W. Cooper, *who participated in the research efforts from early stage, who coauthored one of our previous reports, who passed away as the last pages of the book were being completed.*

Each end is a new beginning! Bill passed away on June 20, 2012 in Texas. On the very same day in Rio de Janeiro, world leaders got together for the United Nations Conference on Sustainable Development (commonly known as Rio+20). A huge crowd of peaceful citizens and militants reaffirmed their commitment to SDGs. With the outcome document *The Future We Want*, the foundation was laid for a new era of partnership for development cooperation. Continuing global deliberations on post-2015 development agenda call for forging a political consensus to fight poverty, inequality, and hunger along with attaining other global SDGs.

History has repeatedly demonstrated that good governance is not a luxury but a vital necessity. Without effective and efficient global governance, it would be impossible to pursue the multifaceted goals of sustainable development, both at the national and global level. Recent crises have taught us that social and economic development is not just about the choice of the right mix of economic and technical inputs but also the effectiveness and fairness of the outputs obtained.

The last decades saw a search for more effective and efficient government to tackle the challenges of globalization in a hierarchically divided world. Recognizing that the national decision-making system does not work well, an increased role must be played by the United Nations, the only supranational institution with a broad political legitimacy. Revitalizing the United Nations and building a new effective system of global governance to support sustainable development will not be easy. Many countries and powerful interest groups have a vested interest in keeping the system as it is, however destructive the results turn out to be, increasing the plight of the poor and depleting even further the scarce resources of the planet. Marshalling opposition beyond the defeatist "after us the deluge" takes time and requires a strong political will.

Yet, as we embark on the Great Reset to build a better post-COVID 2019 World for all, and the *Future We Want*, building on the achievements of the 100 years of the Treaty of Versailles and the operations of the League of Nations, responsible intergenerational choices for the long term become morally imperative. The need for some mechanism of transparent and effective decision-making based on empirical evidence is more obvious than ever.

This new volume after Bill's 100th anniversary is justifiably an occasion for celebration, for recalling his past achievements and for recognizing the contribution that Bill made to operations research and management science (OR/MS). The centenary provides a unique opportunity to consider again the powerful potential of operations research as an analytical underpinning for global decision-making in an increasingly integrated age of globalization. An empirically based and transparent global governance system can aim at the optimal allocation of the world's resources, maximum global welfare, and sustainable development.

In celebration of Bill's 95th birthday, Fred Glover and Toshiyuki Sueyoshi undertook a review of his contributions to OR/MS. As they rightly stated, Bill's research was formative in launching the fields of linear programming, nonlinear programming, goal programming, chance-constrained programming, manpower planning, and multiobjective optimization. Later, he extended the technique of linear programming and nonlinear programming into the development of DEA which has been widely applied to performance analysis in public and private sectors.

The last versions of Cooper's Curriculum, kindly shared by his then assistant Ben Gu, contained 46 pages of most precious information about the life and achievements of this giant of the operational research and management sciences. Bill was a member of about 20 professional societies, and received 55 awards from most prestigious professional associations and institutes:

1945	Co-Recipient (with E. L. Kohler) American Institute of Accountants Award for the year's most significant and valuable article on an accounting subject. (This was the first such award ever given.)
1982	Co-Recipient (with A. Charnes and R. Duffin) of John von Neumann Theory Prize awarded jointly by The Institute of Management Sciences and Operations Research Society of America
1984	Award for Outstanding Research Contributions, Graduate School of Business, The University of Texas at Austin
1986	US Comptroller General Award for Significant Contributions to the US General Accounting Office as a Member of Research and Education Advisory Committee 1970–1986
1987	Award for Research Excellence, Graduate School of Business, The University of Texas at Austin
1988	Distinguished Service to Auditing Award by Auditing Section of American Accounting Association
1988	Auditing Section, American Accounting Association Award for serving as Founding Editor of *Auditing: A Journal of Practice and Theory*
1989	William W. Cooper Fellowships awards initiated by Carnegie Mellon University School of Urban andPublic Affairs (Now the H.J. Heinz, III School of Public Policy and Management)
1991	Notable Contributions to the Accounting Literature Award from the Governmental and Nonprofit Section of the American Accounting Association "for his work in the area of Data Envelopment Analysis"
2002	Lifetime Contributions to Management Accounting Award from the Management Accounting Section of the American Accounting Association. (This was the first such award ever given.)
2002	Robert W. Hamilton Award for Career Research Excellence, University of Texas, May 2002.
2004	Gold Medal Award by The International Society of Multicriteria Decision-making.
2004	Robert L. Mehr award, *Journal of Risk and Insurance*, given by the American Risk and Insurance Association "for that article published 10 years ago that has best withstood the test of time." Coauthored with P.L. Brockett, Linda Golden and Utai Pitaktong.
2006	IMPACT Award by INFORMS (Institute for Operations Research and Management Science).
2006	POMS (Production and Operations Management Society) Award for Contributions to Operations Research and Production and Operations Management.

It is commonly known that professor Cooper's first published article was an economic analysis entitled "The Yardstick of Public Utility Regulation" that appeared in the Journal of Political Economy, June 1943 LI, no. 3, pp. 258–262. But, still earlier in 1938, he published a proceedings paper for the Committee on Capital Gains Taxation of the National Tax Association, which later became an article entitled "Costs, Prices and Profits—Accounting in the War Program," and published in The Accounting Review, July, 1945, val. 20, no. 3, pp. 267–308 along with E.L. Kohler.

His research includes significant contributions to accounting, economics, management, public policy, and other research areas as well as to OR/MS, though we focus here only on his contributions in OR/MS. Throughout his career, Cooper published 28 books and monographs, and 552 articles in leading international journals (Figs. 9.6 and 9.7).

It is a great privilege for us, to be part of Bill's scientific constellation, with a join article that uncovered a new powerful potential of the fractional formulation:

> A Utility Function Approach for Evaluating Country Performances with DEA–The Twin Goal of Decent Work and Fair Globalization." With Sten Thore and Ruzanna Taverdyan in Lee R. Hockley, ed., Global Operations Management (Nova Science Publishers, 2010).

In this last published methodological paper (Cooper et al., 2010), using a logarithmic version DEA, a new interpretation of the fractional programming problem of DEA emerged: maximizing the utility/cost ratio for a single unit to be ranked, subject to the condition that this ratio remain feasible for every other unit.

Our 2010 model, what I would call, Cooper-Thore-Tarverdyan (CTT) Model builds on the original DEA paper (Charnes et al., 1978) that featured a fractional programming model for determination of the efficiency rating of a DMU. Using all statements being in logarithmic units, CTT suggests a multiplicative model as a generalization of the Cobb Douglas (log linear) function. The beauty of this formulation is that it rediscovers the original fractional programming problem and opened the door for the research documented in the present book and much more beyond it.

To Bill, this work reminded him of ideas about Quality of Life pursued by Nestor Terleckyj at the National Planning Association in the early 1970s (Charnes et al., 1973; Terleckyj, 1975). To Terleckyj, Quality of Life was

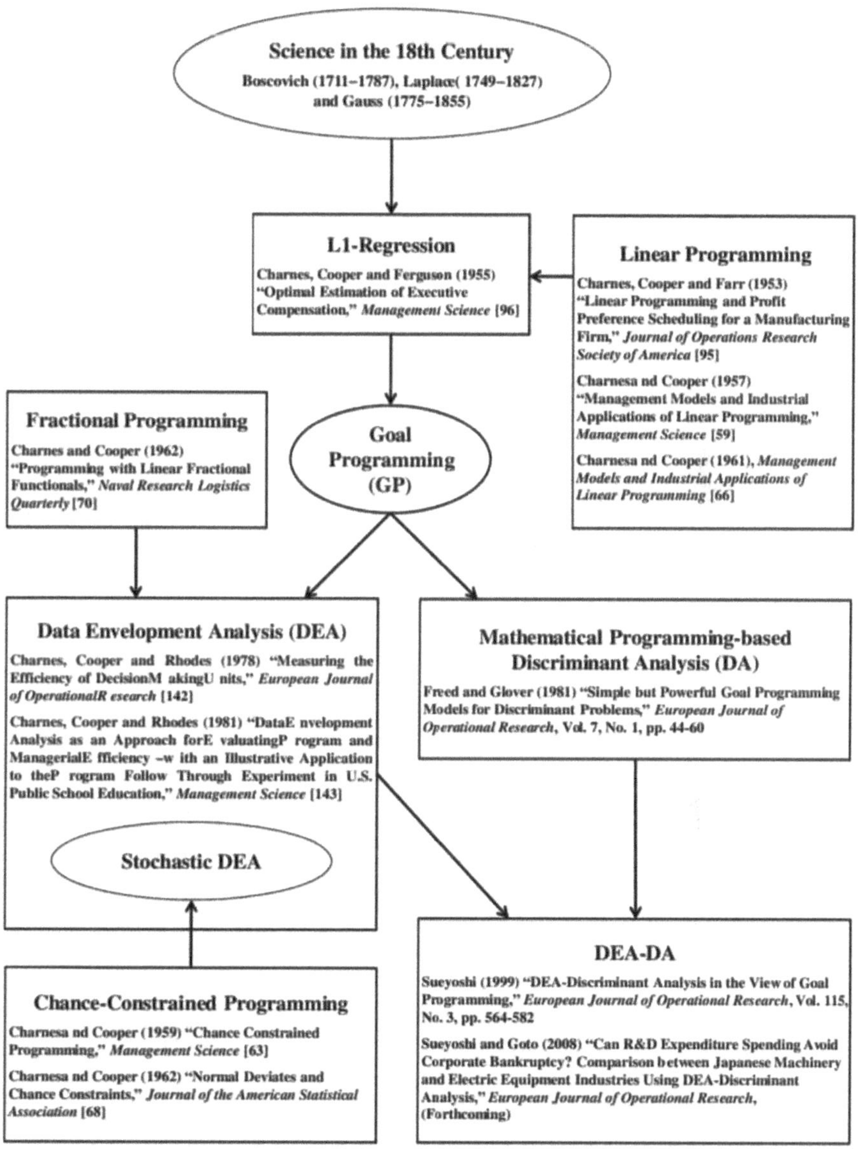

Figure 9.6 Regression, goal programming, and data envelopment analysis.

a multidimensional concept and it should be possible to measure it numerically, using both national account statistics and data from interviews and special studies (Cooper and Ijiri, 1983).

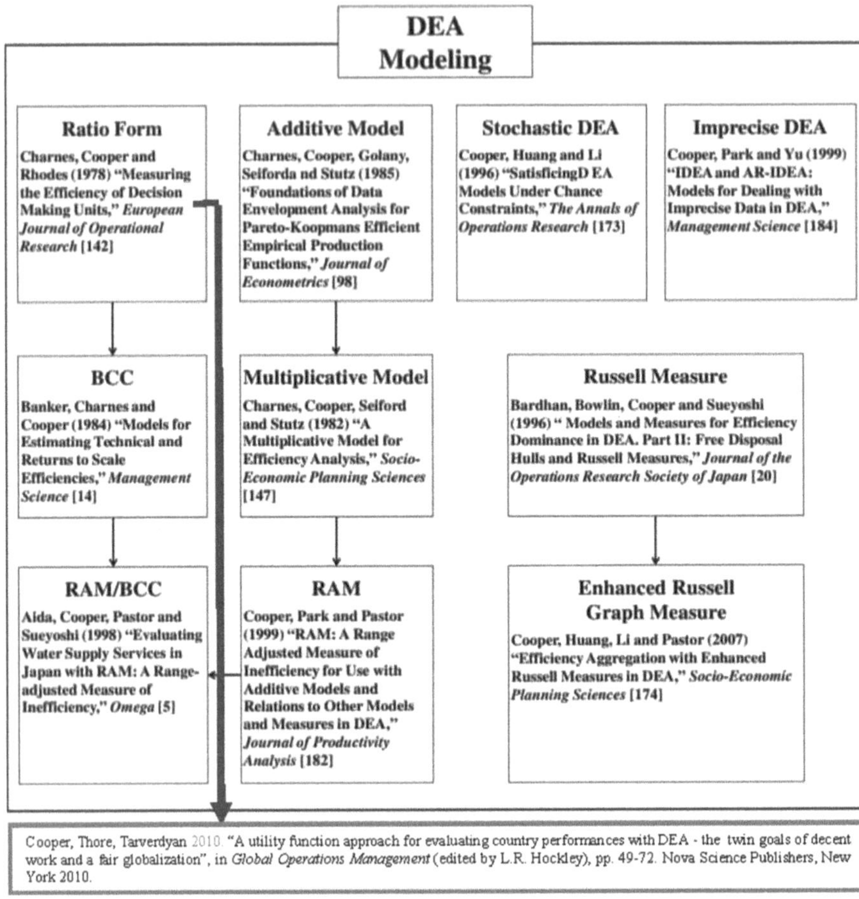

Figure 9.7 Contributions of Professor Cooper in data envelopment analysis models. Source: *Adapted from Glover, F., Sueyoshi, T. European Journal of Operational Research 197 (2009) 1–16.*

Let G_j be a vector of indicators of Goals or the Quality of Life of individual j and $V(G_j)$ is utility function. Let P_j be a vector of policy factors generating Quality of Life and $U(P_j)$ is cost function. The Quality of Life rating of individual j is then defined as the ratio $V(G_j)/U(P_j)$. Maximize this rating for individual $j = 0$, subject to the conditions that the rating can never exceed 1, neither for the individual $j = 0$ himself nor for any other individual j considered:

Fractional CCR 1978

$$\max \mathrm{E} = \frac{\sum_{r=1}^{s} \mu_r y_{ro}}{\sum_{i=1}^{m} \nu_i x_{io}}$$

subject to

$$\frac{\sum_{r=1}^{s} \mu_r y_{rj}}{\sum_{i=1}^{m} \nu_i x_{ij}} \leq 1, \quad j = 1, \ldots, n \tag{9.9a}$$

$$\mu_r, \nu_i \geq 0 \quad \text{all} \quad r \quad \text{and} \quad i,$$

$$0 \leq E \leq 1 \tag{9.9a*}$$

Logarithmic CTT 2010

$$\max \mathrm{QL} = \frac{\sum_{r=1}^{s} \mu_r \breve{G}_{ro}}{\sum_{i=1}^{m} \nu_i \breve{P}_{io}}$$

subject to

$$\frac{\sum_{r=1}^{s} \mu_r \breve{G}_{rj}}{\sum_{i=1}^{m} \nu_i \breve{P}_{ij}} \leq 1, \quad j = 1, \ldots, n \tag{9.9b}$$

$$\mu_r, \nu_i \geq 0 \quad \text{all} \quad r \quad \text{and} \quad i,$$

$$0 \leq GL \leq 1 \tag{9.9b*}$$

where n = number of PMU/DMUs, m = number of inputs, and s = number of outputs. In this notation, $j = o$ represents the DMU_j to be evaluated, which is also included in the constraints Model 9.9a so with $E = 1$ when DMUs is fully efficient relative to the output and input performances of all of the other DMUs included in the $j = 1, \ldots, n$ constraints. Where $\breve{G}_r$, $\breve{P}_i$, and $\breve{G}_{ro}$, $\breve{P}_{ro}$ are logarithmic units for goals and policy variables we have used in our calculations presented in the book.

Goals/SDG indicators	Policy Variables/GDP
$G = (G_1, G_2, \ldots. G_{17})$	$P = (P_1, P_2, \ldots, P_m)$

so $QL = 1$ when PMUs is fully effective relative to the output and input performances of all of the other PMUs included in the $j = 1, \ldots, n$ constraints. Hence, economic and social policy is optimal, generating goals providing the maximal potential sustainability.

In order to determine the effectiveness ratio of one particular country, say country $j = 0$, we form the fractional programming problem. In words, program (Model 9.6a) seeks to determine the virtual weights μ and the constants v and w so that the effectiveness ratio $\mu \log Y_0 / U(X_0)$ for the country currently being evaluated becomes as large as possible, while seeing to it that, using the same weights and constants to calculate the effectiveness ratio of every other country, all these ratios are meaningful, that is, none of them is permitted to exceed 1 (Fig. 9.8 and Models 9.9a*; 9.9b*).

As we have demonstrated the DEA is not just a model but an approach encompassing a large family of models, each model addressed a particular situation and optimization problem making it a powerful tool for problem-solving and decision-making so much longed for by Buchanan.

Considering DEA-derived Pareto optimal solutions as an analytical underpinning for group decision-making toward consensus building, in January 2013, we have constituted the Geneva Consensus Foundation.

As mentioned by Sten Thore, in Diagnostics, "one of the very last projects that occupied Bill's mind was the prolific ideas of our coauthor Ruzanna Tarverdyan, dreaming of creating a foundation in Geneva dedicated to the pursuit of international assessments of national economic and social policy, based on envelopment calculations." I am very happy to have delivered on my promise! The GCF Foundation is a tribute to Bill Cooper, aimed at carrying out his vision of using the DEA in International Decision Making—he so much dreamed of. This is my sacred promise to implement his will, as a token of appreciation for unconditional support I had consistently received from him on my way to Geneva Consensus (his last email was only couple of days before his death).

In 1991, the UN General Assembly adopted the resolution "Restructuring and revitalization of the United Nations in the economic, social, and related fields."[6] The aim of the UN reform thus initiated is to deepen the coherence and enhance the effectiveness and impact of UN development operations, and to intensify the efforts in pursuit of internationally agreed

[6] http://www.un.org/documents/ga/res/45/a45r264.htm.

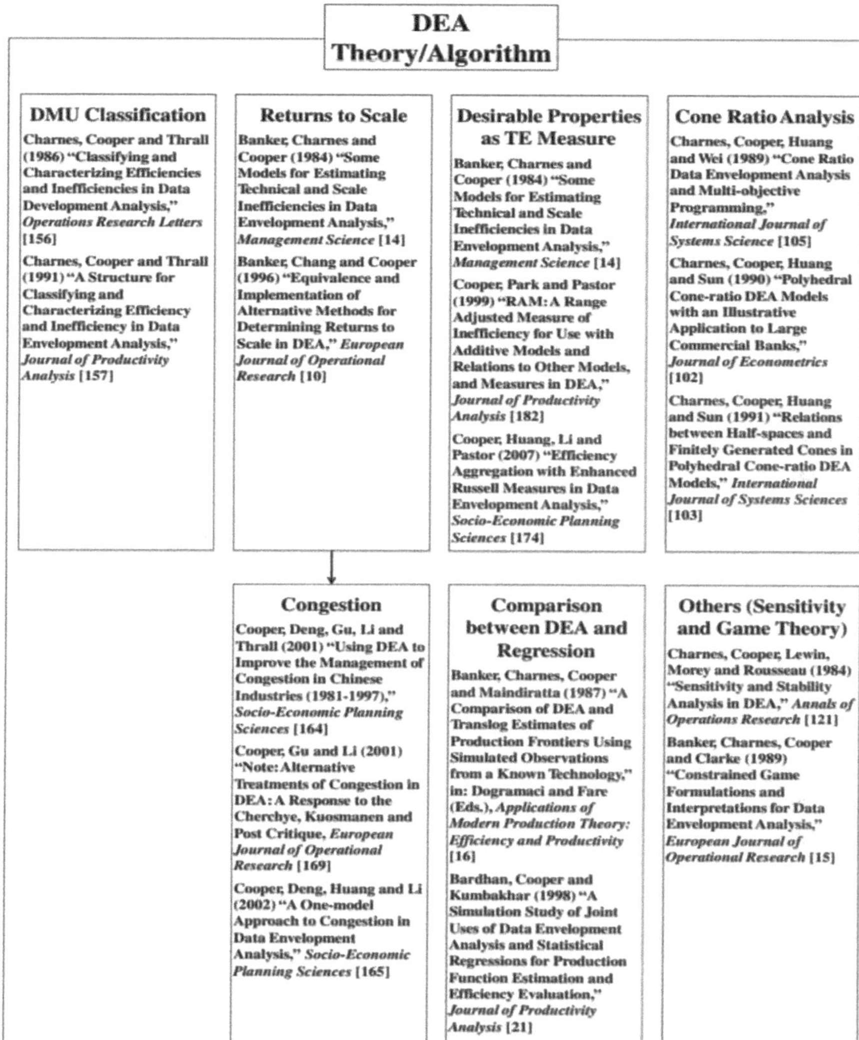

Figure 9.8 Data envelopment analysis theory and algorithms.

development goals. The intention is to make the UN system more result-oriented at the country level. Within the new context, UN operations should be directed toward implementing comprehensive development strategies that can achieve sustainable economic growth and reduce poverty.

Delivering on UN declarations calling for an integrated analysis of SDGs, we propose an *interactive-iterative policy evaluation system for monitoring progress made toward achievement of SDGs*. Our research breaks new ground demonstrating how Tinbergen's grand idea from the 1950s of the optimization of economic policy can be given a contemporary empirical meaning both in a setting of classical equilibrium and for a world in a state of chaotic flux. The operations research technique of "DEA" invented by A. Charnes, W.W. Cooper, and their PhD students has proposed a tool for developing an interactive-iterative monitoring framework for Post-2015 SDGs and implementation of Agenda 2030.

Clearly the international institutional frameworks for sustainable development are poorly developed. Given the multifaceted nature of sustainable development, its attainment will depend upon the ability to form global partnerships drawing on the knowledge and commitment of the WTO, ITC, UNCTAD, UNEP, World Economic Forum, ILO, UNECE, and other UN Agencies and international organizations—all of these actually based in Geneva. The city of Geneva is ideally located to provide a forum for dialog on key policy issues between these institutions and the World Bank, IMF, OECD, and other bimultilateral international organizations.

A new collaborative framework is needed be developed for mobilizing the potential of the world community. The Geneva Consensus Foundation was constituted in January 2013 with the objective:

> **to contribute to World Polity and through the international consensus-building process to put in place a global partnership for sustainable development, to ensure policy coherence for development and for the achievement of Sustainable Development Goals.**

The Geneva Consensus Foundation is a public utility foundation created in compliance with articles 80 et seq of the https://www.admin.ch/opc/en/classified-compilation/19070042/index.html and does not follow any profit or gain.

On January 26, 2016, the United Nations Committee on Nongovernmental Organizations, by the decision ECOSOC/6728-NGO/819 recommended to grant Special Consultative Status to the Geneva Consensus Foundation. At its 16th plenary meeting, on April 5, 2016, the Economic and Social Council of the United Nations adopted a decision 2016/215 to grant consultative status to the GCF. Since 2016 GCF is also a member of the United Nations SDSN.

We submit that the solutions for sustainable development in the interconnected Global world-two multiple dimensional phenomenon, entail a significant paradigm shift toward systems analysis and matrix thinking to address intertemporal effectiveness-efficiency-equity challenges for future generations. We believe Rio+20 and Agenda 2030 have laid solid foundations for a paradigm shift. Unfortunately the lack of integrated systems thinking in most International/UN organizations prevents identification of solutions to contemporary global challenges.

Our foundation, an infant think tank, aims to bridge this gap and we truly hope we will be able to make our modest contribution to global deliberations in shaping a global agenda to harness the power of globalization to advance inclusive, just and sustainable development, for enduring peace and progress.

We believe that the Operations Research/Management Science (OR/MS) offer analytical tools that have a great potential for application at the global level. Rio+20 has already laid solid foundations for a paradigm shift and the introduction of the new SDGs agreed during the 69th General Assembly in September 2015 and the Agenda 2030 promis to encourage the construction of global operations research models aiming at *the optimal use of the world's resources in accordance with the objective of sustainable development.*

Unfortunately there is a widespread lack of global modeling skills in the UN and most international organizations that prevents the identification of optimal solutions to contemporary global challenges. The Geneva Consensus Foundation aims to bridge this gap *by providing analytical underpinnings that the global decision-makers need*, enabling them to make evidence-based decisions to advance inclusive, just, and sustainable development for enduring peace and progress. The recently founded *Geneva Consensus Foundation* is the kind of independent institution that could identify opportunities and mechanisms for possibly improved performance and fostering synergies toward key policy issues.

9.9 Geneva Consensus global decision-making system

From an analytical perspective leaving no one behind and attainment of SDGs makes a case of looking for Pareto-Koopmans efficient solutions. What an economist would call the Pareto principle a political theorist might call the consensus principle. With our research expanded over the past 15 years, as a first and modest step toward developing *a new economic theory* more adept to solve the contemporary problems facing our COVID crisis hit global world. Here we unveil a marvelous power of the interconnected sets of DEA models. Much inspired by the critics and hopes expressed by a number of Nobel Memorial Prize Laureates in Economics, presented in earlier chapters of this book, we propose a scientifically supported decision-making system capable of measuring effectiveness, efficiency, and equity of multidimensional performances of various units that comes to fill the persistent intellectual gap among various economics schools.

An approach, based on the family of DEA Models, which has been tested over 40 years, and can be applied in various contexts and levels: ranging from households to governments, from local to global levels. We thus propose a new consensus building framework that will enable to evaluate how various decision and policy-makers succeed (or fail) in their starvings for improving their transformation processes and the betterment of the lives of their future generations in the age of Global post-COVID Reset.

In performance evaluation exercise, the research can only be viewed as fulfilling a part of the process a complete evaluation would rely on the interaction between research, policy-making, and institutional development. Bringing about possible positive changes will entail policy action. This will require systematic efforts in the areas of policy planning, implementing, and controlling. And as mentioned in Preface to "Diagnostics for a Globalized World," by Sten Thore,

> This book examines the promise in this regard that is offered by data envelopment analysis (DEA), a quantitative technique invented by A. Charnes and W.W. (Bill) Cooper and their PhD students. This promise is documented in publications spanning a twenty-year period and in the chapters to follow—but the large-scale practical application is still awaiting. **To prepare for this practical phase, my coauthor Ruzanna Tarverdyan has established the foundation The Geneva Consensus which operates out of Geneva, Switzerland.**

The evaluation of national performance can only be part of the process, however. The value of any empirical research is greatly limited to the further use suggested in findings and their policy impact. Ensuring model-driven improvements would rely on the interaction between research, policy-making, and institutional development. Further, bringing about possible positive change will entail policy action and requires systematic efforts to assess and analyze *policy interventions* on the international scene. The GCF, that enjoys Special Consultative Static with the UN ECOSOC since 2016, with a recognized competence in sustainable development domain, in compliance with its constitutional objective, is dedicated to the promotion of DEA-based research in UN deliberations, on the path of implementation of Agenda 2030 for the achievement of the SDGs.

Within this context, consensus building is particularly important for reconciling the range of potential interests in various public development domains. In the framework of cooperative negotiations, finding mutually acceptable solutions will promote acceptance and support for the agreements arrived at and thus ensure larger ownership over the agreed measures and their smooth implementation. To assess and analyze policy interventions on the international scene, and identify opportunities and mechanisms for possibly improved performance and foster synergies toward key policy issues, a new collaborative framework shall be developed for mobilizing the potential of the world community.

Mechanisms for participatory decision-making processes that will manage effectiveness-efficiency-equity trade-offs in globalization and will enable all development stakeholders to make informed choices and build a consensus on managing the process of globalization and plan future development policies of countries toward the achievement of the SDGs need to be developed. We submit that, only within a new global partnership will the global world be able to strike a balance and create the foundations for innovative policy packages ensuring convergence of the well-being of people, planet, and societal prosperity:

- Understandings between developing and developed countries play a key role, both within country groups as well as in global society,
- Through regional and global dialog, a more inclusive form of globalization can emerge.

Consensus is the foundation of stability and sustainability, forging an international bipolar, triangular, and global consensus is a potent mechanism for shaping national and international policies that are socially just, economically productive and environmentally friendly.

Delivering on UN declarations on integrated analysis of multiple SDGs, and Agenda 2030 for Sustainable Development, we developed a DEA-based policy evaluation system for monitoring progress made by various PMUs and DMUs, toward achievement of SDGs to ensure policy coherence. It is an *interactive* and an *iterative* input-output framework that enables to *elicit* maximum level of consensus, based on the estimates of relative values in the best interest of stakeholders. Our approach is consistent with empirical data, rather than imposing assumptions a priori.

As already noted, the DEA is a benchmarking tool for evaluating the performance of a population of entities called DMUs—enterprises, regions, etc., aiming to find the best production/transformation practices that has received a wide application during the past 40 years. DMUs are responsible for converting input to outputs. The ultimate goal is to improve these processes of inefficient DMUs given their mix of inputs and outputs in comparison with the DMUs on the frontier with a similar mix of indicators. The DEA was originally developed by Charnes et al. (1978) solely for the purpose of evaluation of efficiency to ensure the best achievable set of outputs for a given set of inputs, but was then expanded to effectiveness considerations. In our research, we shifted from DMUs to a notion of PMUs and evaluated the effectiveness of countries' policy performance toward development goals.

The large and growing number of applications of the DEA model is explained by its distinguishing feature to handle multiple inputs as well as outputs that are interacting in complex and unknown ways and may be measured in different units. The DEA is a "black-box" technique used when there is insufficient knowledge of causal relationship, and it is difficult to ascertain or test it exactly; or there is a lack of consensus on the alternatives. It provides useful results for conducting *ex-post* evaluation of performance of DMUs that enables decision makers to agree upon ex-ante measures toward improving their performance.

Within the DEA context, when evaluating the performance of each PMU/DMU, mathematical programming model is applied to the input-output data for *all* PMU/DMUs in order to evaluate the performance of *each* PMU/DMU. Thus the DEA characterizes each PMU/DMU by a single summary relative efficiency score. The fractional formulation of the model, using *the ratio of the weighted sum of outputs to the weighted sum of inputs,* provides a contact between the efficiency concept of welfare economics and of engineering science.

To better *link development outcomes to government policies (and donors' interventions)* and measure the efficiency/effectiveness/equity of government policy implementation in the context of globalization, we suggest a new input-output paradigm of DEA-based Global Decision Making System (Table 9.3):

Reverting back to what Buchanan submitted in his contribution to "The State of Economic Science: Views of Six Nobel Laureates":

> *If the political economy is conceived as being described, in part, by constraints that can be subject to explicit collective choice, attention is immediately drawn to prospects for constitutional-institutional change. Once again the game analogy is helpful; we change a game by changing the rules, which will, in turn, modify the predicted pattern of outcomes. If we diagnose the patterns of results observed to be less desired than alternative patterns deemed to be possible, it is incumbent on us, as political economists, to examine predicted results under alternative constraint structures.*

Table 9.3 Inputs and outputs in Data envelopment analysis.

INPUTS	Inputs	Costs	Undesirables	Risks	Bads	Efforts	*Policy variables*
OUTPUTS	Outputs	Benefits	Desirable	Returns	Goods	Satisfaction	*Performance variables*

- Inputs and outputs should be related experientially, statistically and/or conceptually, and it is also important to have information on the direction of the relationship. Whether it is positive or negative.
- Depending on data availability, surrogates might be used.
- Additional extensions could proceed to a two-stage operation in which development "outputs" obtained from the inputs in stage 1 serve as inputs in stage 2 and are translated into final desired "outcomes."
- The two-stage DEA model exploits the value of decomposing the analysis of an activity into two or more linked models.

In accordance with DEA terminology "satisfactions," Goals can be considered as outputs and "efforts" as inputs. The ratio "satisfaction to efforts" could serve as a measure of the efficiency of the unit Elias Maragos and Dimitris Despotis, 2002 "Comparing multiobjective mathematical programming methods in the light of Data Envelopment Analysis," Department of Informatics, University of Piraeus, Greece

- ***Outputs/outcomes/goals*** are the objectives a country or a society wants to achieve, which are often expressed in nontechnical, qualitative terms;
- ***Indicators*** are the variables used to measure progress toward the goals;
- ***Targets*** are the quantified levels of the indicators, that a country, or society, wants to achieve at a given point in time, the SDGs are a good example for that;
- ***Inputs*** are policies implemented to achieve the above outputs/goals.

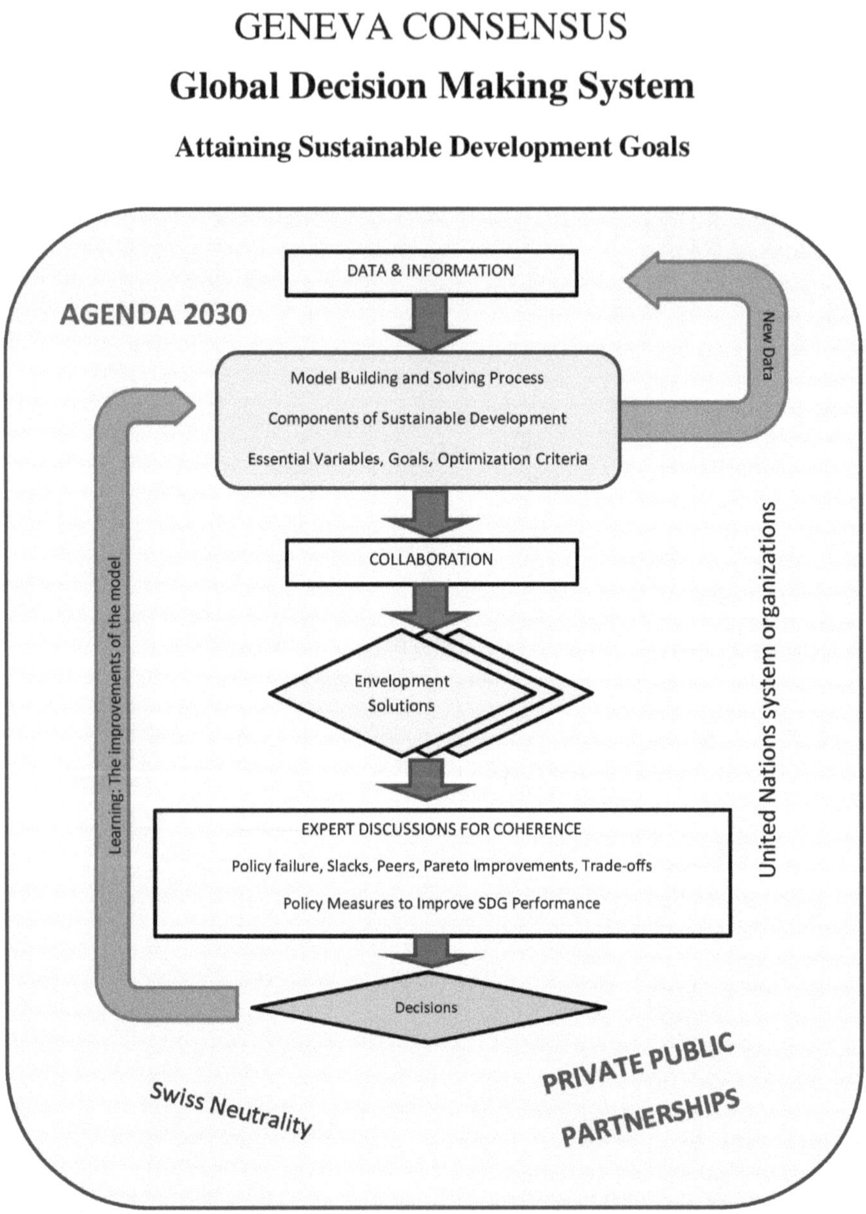

Figure 9.9 Geneva consensus framework: Pareto principle and consensus principle.

Recognizing the above-mentioned properties of the DEA models, we propose *DEA* (Fig. 9.9) as the:

- participatory, interactive-iterative performance evaluation system that expands the notion of competitiveness of nations, businesses beyond the economic dimension,
- basis for integrated packages of policy advice for comprehensive global, regional, and industry agendas,
- surveillance instrument for global governance to track development progress; to achieve consensus on how to simultaneously maximize benefits and minimize the costs of globalization.

The construction of typical DEA model will consist of the following elements:

1. definition of PMU/DMUs,
2. input-output variables, *discretionary or nondiscretionary variables, categorical—dummy variables,*
3. choice of orientation and a scale assumption—*input minimization or output maximization, constant returns to scale or variable returns to scale*
4. weight restriction: *judgment—managerial preferences or expert opinion, incorporation of known standards* longitudinal analysis.

In DEA, for each PMU/DMU, we deal with multiple inputs and outputs which are assigned weights (to be determined from the data) with the following model:

$$\max \mathrm{E} = \frac{\sum_{r=1}^{s} \mu_r y_{ro}}{\sum_{i=1}^{m} \nu_i x_{io}}$$

subject to

$$\frac{\sum_{r=1}^{s} \mu_r y_{rj}}{\sum_{i=1}^{m} \nu_i x_{ij}} \leq 1, \quad j = 1, \ldots, n$$

$$\mu_r, \nu_i \geq 0 \text{ all } r \text{ and } i, \tag{9.10}$$

where n = number of PMU/DMUs, m = number of inputs, and s = number of outputs. In this notation, $j = o$ represents the DMU_j to be evaluated, which is also included in the constraints so:

$$0 \leq E \leq 1, \tag{9.11}$$

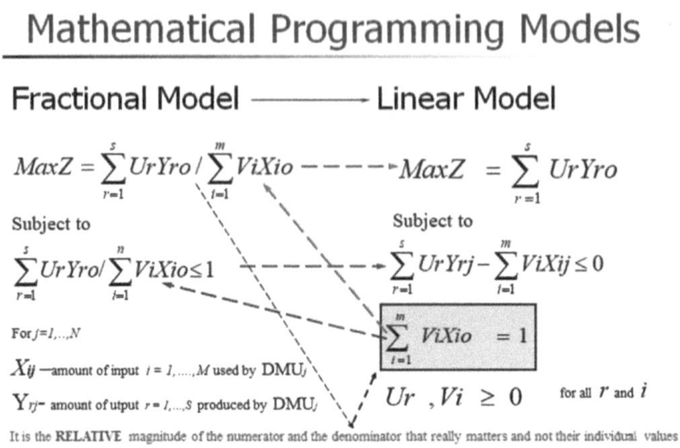

with $E = 1$ when DMUs is fully efficient relative to the output and input performances of all of the other DMUs included in the $j = 1, \ldots, n$ the constraints. The μ_r and ν_i which are to be determined are referred to as "multipliers," in order to distinguish them from the customary uses of predetermined weights, and the sums.

$\sum_{r=1}^{s} \mu_r^* y_{ro}$ and $\sum_{i=1}^{m} \nu_i^* x_{io}$ are called "virtual outputs" and "virtual inputs," respectively, with "*" indicating optimal values.

Problem (Model 9.10) is nonlinear and difficult to solve. However, Charnes and Cooper (1962) showed how this problem, referred to as a "fractional programming problem," can be transformed into an ordinary linear programming problem that is readily solved by any of numerous available linear programming computer codes.

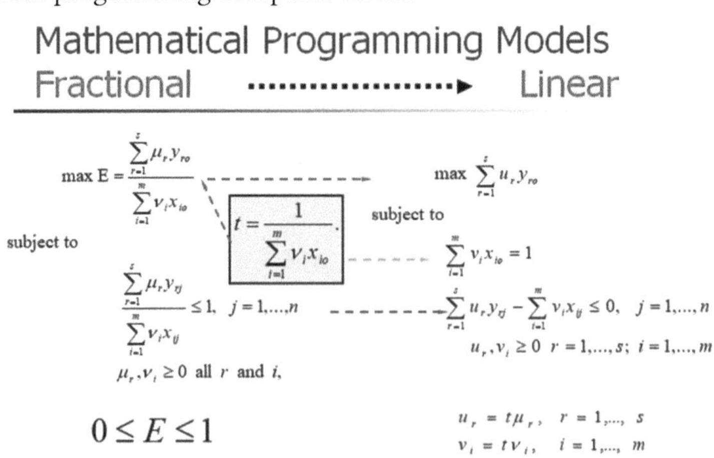

CHARNES COOPER TRANSFORMATION 1962

To see how this is done, we define a new "transforming variable" by:

$$t = \frac{1}{\sum_{i=1}^{m} v_i x_{io}}. \tag{9.12}$$

Multiplying numerators and denominators in both the objective and the constraints by t, leaves their values unchanged. Then we can replace the variables μ_r and ν_i with new variables defined by:

$$\begin{aligned} u_r &= t\mu_r, \quad r = 1, \ldots, s \\ v_i &= t\nu_i, \quad i = 1, \ldots, m \end{aligned} \tag{9.13}$$

so, utilizing these new variables, we can replace (1) with:

$$\max \sum_{r=1}^{s} u_r y_{ro}$$

subject to

$$\begin{aligned} &\sum_{i=1}^{m} v_i x_{io} = 1 \\ &\sum_{r=1}^{s} u_r y_{rj} - \sum_{i=1}^{m} v_i x_{ij} \leq 0, \quad j = 1, \ldots, n \\ &\quad u_r, v_i \geq 0 \quad r = 1, \ldots, s; \quad i = 1, \ldots, m. \end{aligned} \tag{9.14}$$

As shown in Charnes and Cooper (1962), problem (Model 9.14) has the same optimal value as (Model 9.10). Moreover the first constraint in (Model 9.14) guarantees that not all v_i can be zero. This, in turn, guarantees that $t^* > 0$ so solutions to (2) may be used to obtain solutions to (Model 9.10) by dividing by t* and reversing the transformations in (Model 9.13).

Since (Model 9.14) is a linear programming problem, it has a dual problem formed from the same data, which we write as follows:

$$\min \theta$$

subject to

$$\begin{aligned} &\sum_{j=1}^{n} x_{ij}\lambda_j \leq \theta x_{io}, \quad i = 1, \ldots, m \\ &\sum_{j=1}^{n} y_{rj}\lambda_j \geq y_{ro}, \quad r = 1, \ldots, s \\ &\lambda_j \geq 0, \quad j = 1, \ldots, n. \end{aligned} \tag{9.15}$$

The optimal values of Models (9.14) and (9.15) are equal so that:

$$\min\ \theta = \theta^* = \sum_{r=1}^{s} \mu_r^* y_{ro} = \max \sum_{r=1}^{s} \mu_r y_{ro} \tag{9.16}$$

Dual Problem and "Pareto-Koopmans" efficiency

$\min\theta$

subject to

$$\sum_{j=1}^{n} x_{ij}\lambda_j \le \theta x_{io},\ i=1,\ldots,m$$

$$\sum_{j=1}^{n} y_{rj}\lambda_j \ge y_{ro},\ r=1,\ldots,s$$

$$\lambda_j \ge 0,\ j=1,\ldots,n$$

$$\min\theta = \theta^* = \sum_{r=1}^{s} \mu_r^* y_{ro} = \max\sum_{r=1}^{s} \mu_r y_{ro}$$

$$E^* = \sum_{r=1}^{s} \mu_r^* y_{ro} = \theta^*$$

It follows that:

$$E^* = \sum_{r=1}^{s} \mu_r^* y_{ro} = \theta^* \tag{9.17}$$

The dual pair of problem operates in the following manner: One problem locates inefficiencies and their amounts in the performances of each DMU. The other problem evaluates the contribution to the value of the overall measure of inefficiency for each input and each output. It is not necessary to solve the two problems separately since the computer codes simultaneously supply solutions to both: *Optimize both by maximizing outputs and minimizing inputs simultaneously.*

Examination of (Model 9.15) shows that no input or output can be improved at the expense of any other input or output. This leads to the following definition of efficiency, called the "Pareto-Koopmans" definition of efficiency since it corresponds to the definition of "Pareto efficiency" that is used in welfare economics, See David. W. Pearce (1980).

Definition 1: Efficiency is achieved if and only if $\theta^* = 1$ *which means that it is not possible to improve any input or output without worsening some other input or output.*

Via (Model 9.17) this means that $E^* = 1$ so DMU_o, the DMU being evaluated, is positioned on the frontier of the "production possibility set" generated by DEA from the x_{ij} and y_{rj} that serve as the coordinates of the observed inputs and outputs for each $DMU_j, j = 1, \ldots, n$.

The following definition of inefficiency then also applies.

Definition 2: Inefficiency is present when it is possible to improve some input or output without worsening some other input or output.

Notice that these definitions eliminate the need for assuming values that supposedly represent degrees of relative "worth" or "value" between the inputs and outputs. Notice also that this condition is not present in (Model 9.10), which can be regarded as a generalization of the one-input-to-one-output ratio definition of efficiency used in engineering and science.

The two definitions (engineering science and Pareto-Koopmans) are connected, however, by the duality relations described above. Moreover, either definition may be used without losing contact with the other since linear programming (and DEA) computer codes generally supply solutions to both problems when solving either of the above dual problems. For further discussion, see Cooper et al. (2005).

Key constructs of a DEA are the envelopment surface and the efficient projection path to the envelopment surface. The DEA proceeds by constructing an empirical surface (frontier-benchmark) composed of the best practice performers and then measures the efficiency relative to that frontier.

DEA identified inefficiencies for those DMUs not on the frontier providing the amount of additional output that could have been produced if the DMU in question operated using the "technique" of the best practice provider. And identifying the group of peers the DEA enables to learn from best practice frontier. DEA also provides reference units for inefficient DMUs. Reference units are hypothetical units on the efficient surface, which can be regarded as target units for inefficient units. A reference unit is traditionally found in DEA by projecting an inefficient DMU radially to the efficient surface (Fig. 9.4). Term "radial" means that an efficient frontier is tried reach by incorporating the two values of the current outputs and/or decreasing the values of the current inputs in the same proportion (Fig. 9.10).

Recognizing that there exist no absolute performance evaluation systems, we propose the DEA to become an *interactive-iterative* policy evaluation system, enabling to *elicit* maximum levels of consensus. These are to be based on the estimates of relative values in the best interest of stakeholders and consistent with country data, rather than a *priori impositions*.

This would allow the analyst and main stakeholders to represent different country subgroups and to study model results in view of their realities. This could suggest additional preference information that could further be incorporated into the following phases of the analysis.

The model results might provide early warning about imminent development trends and could serve as a basis for the development of an integrated package of policy advice helping to track progress toward *selected*

Envelopment Procedure

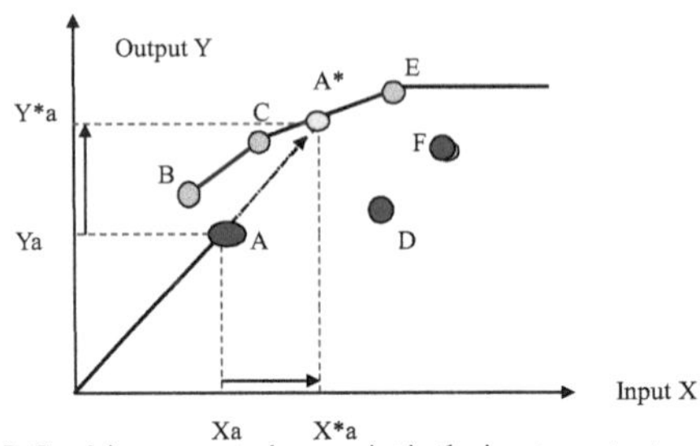

Each country (A,B,C,....) is represented as a point in the input—output space.
The envelope B-C-E is the upper convex and non-decreasing *hull* that can be fitted to the points
Each unit not on the frontier is scaled against a convex combination of units on the frontier facet closest to it.
DEA is sensitive to extreme values and hence some interesting but unusual outliers that lack comparisons can be picked and singled out by the model.

Figure 9.10 Understanding the envelopment procedure.

policy goals. They could also become a strategic means of institutional support to UN member states, which could be done in collaboration with other development institutions and other stakeholders, toward the achievement of benchmark goals of SDGs, or other globally established goals and ultimately ensuring sustainable development.

As objectively criticized by Buchanan, standard calculations of GDP provide no clue as to how a nation could possibly improve its performance. They meekly record actual achievements, whether in the upper range or the lower range. The great advantage of the effectiveness rating of policy to be defined here is that it is *diagnostic*: it holds specific suggestions as to how the sustainability achievements of the nation could be improved toward Pareto optimally.

Our research is founded on the fact that there is no a natural trade-off for a nation between various dimensions of sustainable development, and that it is a matter of policy choices and prioritization in the framework of given strategic development goals of countries. Our framework offers the possibility for establishing such trade-off decisions, and, using the vocabulary of Buchanan, the wide application of "exchanges" decisions, at various levels and contexts, will be possible, once the necessary data become available (United Nation (2013a); United Nations (2015)).

By the virtue of the DEA Dual Models, the endogenously selected weights maximize the weighted sum of index value of the country performance and provide information that can be used to guide *trade-offs* between different inputs/outputs or to determine the results of increases/decreases in their

DEA FRAMEWORK

Task:	Understand	Assess& Evaluate	Measure	Compare	*the*
Observables:	Business practices	Policies, institutions	Transformation processes	Operations : inputs and outputs	*to*
Objectives:	Perform Country Comparisons	Improve country performance	Resign regulatory and institutional incentives	Establish policy priorities	*Given the needs of*
Context:	Management	Stakeholders	Reform Globalization	Public policy	

Figure 9.11 Benchmarking menu. Source: *Adapted from Paralaz, L. L (1999). Utility Benchmarking on the West Coast.* Journal of the American Water Works Association.

corresponding amounts. One of the prominent features of original DEA models is the fact that it makes each DMU to be seen in the best possible light compared to others. The DEA approach perhaps is the easiest and offers the most promising and flexible method to determine capacity and capacity utilization. The optimal values of the coefficients or weights (called multipliers) represented by the U_r^* and V_i^* represent the rate of contribution per unit of the corresponding output or input amount to the performance score, h_o^*, for DMU_0. These U_r^* and V_i^* provide information that can be used to guide trade-offs between different inputs and outputs or to determine the results of increases and decreases in their corresponding amounts.

In the competitive style of negotiation, each participant is following the objective to win, in a sense of doing better, or getting more, than the other negotiating parties. One last, yet a very important feature of our proposed DEA framework lies in its capacity to ensure empirically derived solutions satisfactory to all PMUs/DMUs under consideration, to serve as an analytical ground for forging *team consensus* within the context of Groupthink. Moreover given the state of Pareto optimality, it guarantees a win-win solution for all negotiating parties. Building on the work of Paralaz. L (1999), we propose a DEA Framework for setting up the objectives and delibering on (Fig. 9.11).

Framework Linking Policies and Actions with Development Outcomes

Policies and actions			Intermediate outcome	MDGs and SDGs
Economic and Financial Polices • Macro • Trade • Private sector regulatory & Institutional environment • Infrastructure: Physical & Financial	***Public Sector Governance*** • Public expenditure & revenue management • Public administration • Transparency, accountability, control of corruption	**Developing Countries** • create a supportive enabling climate for economic activity • empower and invest in poor people	**STRONGER AND SUSTAINABLE ECONOMIC GROWTH**	**MDGs SDGs**
Social sector • Building human resources: education and health • Social protection • Voice inclusion • Gender	***Environment*** • Policies and Institutions			
Trade • Market access(including domestic subsidies) • Trade-related assistance	***Aid*** • Quality • Quantity	**Developed Countries** • Enhance support to developing country reform efforts and global public goods	**IMPROVED DELIVERY OF SERVICES TO POOR PEOPLE**	
Debt Relief • Enhances HIPC initiative	***Overall Macro Policies*** • Stable & growing world economy			

Source: Adopted from: Achieving the MDGs and Related Outcomes: A Framework for Monitoring Policies and Actions, the World Bank and the IMF Development Committee, March 26, 2003

SDGs have made a qualitative leap forwards since the establishment of the Millennium Goals (Adams and Judd, 2018). The differential factor of greatest interest here is probably the inclusion in the evaluation and advancement process of all the world's countries, not just developing ones. The objectives, targets, and indicators highlight not only deficits in development (poverty, water sanitation, access to food) but also issues such as those related to the misuse of resources and the pollution of ecosystems, where more developed countries tend to perform worse.

UNDP and World Bank. Transitioning from the MDGs to the SDGs; UNDP: New York, NY, USA; World Bank: New York, NY, USA, 2016

Matrix 1: The Sustainable Development Framework

SDF	The Prerequisites for Sustainable Growth and Poverty Alleviation													
	Structural				Human		Physical				Specific Strategies			
The Activities of Partners in the Development Process	*(1)* Good and Clean Governance	*(2)* Justice System	*(3)* Financial System	*(4)* Social Safety Net and Social Programs	*(5)* Education	*(6)* Health And Population	*(7)* Water and Sewerage	*(8)* Power	*(9)* Roads, Transportation and Telecomunications	*(10)* Environmental and Cultural Issues	*(11)* Rural Strategy	*(12)* Urban Strategy	*(13)* Private Sector Strategy	*(14)* (Country Specific Headings)
Government National Provincial Local														
Multilateral and Bilateral Institutions														
Civil Society														
Private Sector														

9.10 Data revolution

In the earlier parts of this book, we have demonstrated empirically how is it possible to pursue a policy of economic growth and at the same time reach for the developmental and environmental goals pronounced by the *UN Agenda 2030*! UN agencies are now involved in the construction of a data repository on 17 SDGs. This is a huge task that will no doubt take considerable time and significant resources.

Gregory Bateson in Form, Substance and Difference published in *Steps to an Ecology of Mind (Bateson 1972)*, in 1972 defined the notion of "information" or "negative entropy":

> The technical term *"information"* may be succinctly defined as any difference which makes a difference in some later event.
>
> what we mean by information—the elementary unit of information—***is a difference which makes a difference***,[7]

The Secretary-General's High-Level Panel on the Post-2015 Development Agenda reiterates the vital importance of building data systems to provide timely, disaggregated indicators to measure progress, in all countries, and at all levels (local, subnational, and national).

The availability of statistical information has considerably improved after the adoption of the SDGs. Learning from data analysis—and adapting actions based on what we learn from evaluation—is one of the best ways to ensure that development goals are reached. The revolution in information and communication technologies over the last decade provides an opportunity to strengthen data collection and statistical analysis for accountability and decision-making purposes. Mobile technology other innovative initiatives and advances have also been used and to enable collection of real-time data.

Better data and statistics will be a considerable help to governments to track progress and make sure their decisions are evidence based. They can also strengthen accountability not just for national governments but also international agencies, CSOs, and the private sector should be involved. A true data revolution would draw on existing and new sources of data and will

[7] Bateson, G. "Steps to an Ecology of Mind: Collected Essays in Anthropology, Psychiatry, Evolution, and Epistemology." San Francisco: Chandler Pub. Co, 1972.

fully integrate statistics into decision-making, promote open access to, and use of, data and ensure increased support for statistical systems.

Toward this end, the report[8] of the Secretary-General's High-Level Panel on the Post-2015 Development Agenda entitled "A New Global Partnership: Eradicate Poverty and Transform Economies through Sustainable Development," released in 2013 recommended establishing a Global Partnership on Development Data that will bring together diverse but interested stakeholders—government statistical offices, international organizations, CSOs, foundations, and the private sector.

This partnership would, as a first step, develop a global strategy to fill critical information gaps, expand data accessibility, and galvanize international efforts to ensure a baseline for informed decision-making for the 2030 targets. Investments in data gathering will be needed to ensure data for indicators that are not yet available. Technical experts meetings are foreseen to discuss and refine the models and methods for qualitative indicators that are not yet agreed or are unclear.

DEA co-founded by Bill Cooper provides a strong foundation for integrating discipline-based knowledge into a single analytical framework that can become an analytical tool for developing more specific, operationalizable global governance models. Within the formal structures of global governance, negotiations and consensus building processes DEA-based models will yield optimal solution for the global world, going well beyond global economy to cover the challenges facing the environment, global society, etc.

The DEA provides a coherent cycle of data analysis performance enhancement quiet harmonious with information-feedback-behavioral change.

This interesting feature of envelopment analysis originated in 1978 and explains its wide-scale practical application to enhance performance of DMU of totally different nature. The DEA-based performance evaluations have culminated in the expanded version of its application to measuring policy effectiveness of countries as presented in the part one of the book.

Moreover, the DEA model solutions, containing the optimal levels of variables, slakes, efficiency losses, identification of peers, etc., connect perfectly well with the cybernetics concept of learning as an indirect method of regulation, and the adoption of a pattern of behavior that is compatible

[8] http://www.post2015hlp.org/the-report/.

with survival. This very exceptional advantage of DEA: to account for improvements and technological chance, was long disregarded in earlier analytical frameworks and economic theories.

The Marrakesh Agreement establishing the World Trade Organization states:

> *Recognizing that their relations in the field of trade and economic endeavour should be conducted with a view to raising standards of living, ensuring full employment and a large and steadily growing volume of real income and effective allowing for the optimal use of the world's resources in accordance with the objective of sustainable development ….*

In fact the establishment of GATS/WTO and its continuous deliberation toward creation of level playing field is a manifestation of global political will to reduce the differences among the territories to be reflected on the global map.

Consequently when looking with cybernetic lenses at the standard setting and normative deliberations of the UN System, they should be viewed as manifestation of enabling environment for reducing the variety in the disturbances in the system of the globalized world, hence contributing to a decrease in entropy that in turn will increase the predictability in the global system and serve as a guarantee for global stability. From its inception the GATT/WTO was built on the practice of consensus decision-making, the major challenge remains however on how to prevent the consensus building process from unjustified exercise of veto without giving a reasoned explanation for their opposition.

The proposed DEA-based "Geneva Consensus" framework provides a solution to these problems in view of the fact that Pareto principle might act as a consensus principle. It is grounded in the principle of collaborative consensus building among already WTO 159 member countries toward DEA-based win-win solutions. Based on cybernetic feedback processes and informed evidence-based negotiations, mutual interests could be accommodated. The proposed governance framework will enable to effectively use the institutional capacity of the WTO and other international organizations and ensure that neither issues opposed by many are pushed, nor progress desired by many blocked.

Enhanced data availability will create the information basis for development of DEA-based global cybernetic models, thus providing an analytical underpinning for global governance.

The advent of information and communication technologies, the methodological advancement of statistical development and analysis, and most importantly the "data revolution" called by the report of the UN Secretary-General's High-Level Panel[9] will create legal, political, and methodological grounds for such a change.

Recognizing that there exist no absolute performance evaluation systems, we propose the DEA to become an *interactive* and an *iterative* policy evaluation system, enabling to *elicit* maximum levels of consensus. These are to be based on the estimates of relative values in the best interest of stakeholders and consistent with country data, rather than a *priori impositions*. This would allow the analyst and main stakeholders to represent different country subgroups and to study model results in view of their realities. This could suggest additional preference information that could further be incorporated into the following phases of the analysis.

The model results might provide early warning about imminent development trends and could serve as a basis for the development of an integrated package of policy advice helping to track progress toward *selected* policy goals. They could also become a strategic means of institutional support to UN member states, which could be done in collaboration with other development institutions and other stakeholders, toward the achievement of benchmark goals of sustainable development or other globally established goals and ultimately ensuring sustainable development.

9.11 Way forward: a quest for a new paradigm

Our preliminary results in evaluating the integrated social, economic and environmental performance of countries with the application of the DEA, documented in multiple scenarios, empirically confirm that there exists a huge Sustainable Development Deficit, the world is still far from realizing the full potential of globalization. However, *if* wisely ***managed***, globalization *can* deliver unprecedented material progress, generate more productive and better jobs, and contribute significantly to reducing poverty and improving the quality of life. We thus, find that:

- a better world is *possible*
- the benefits of globalization *can* be extended to *more* people and countries
- the potential for good is *immense* and *can* be achieved in the context of *open* economies.

[9] http://www.post2015hlp.org/the-report/.

Figure 9.12 Policies, institutions, sustainable development.

To better *link development outcomes to government policies (and donors' interventions)* and measure the efficiency/effectiveness/equity of government policy implementation in the context of globalization, we suggest a new input-output paradigm of DEA (Fig. 9.12):

Inputs being indicators of national development *policies*: macroeconomic, fiscal, social, monetary, trade, aid, migration, institutional-public and corporate governance, etc., and

Outputs being *performance* indicators for SDGs, targets: social, economic, environmental, sustainability (Fig. 9.13).

The DEA score will demonstrate efficiency/effectiveness of policies (Fig. 9.13) and will:

- evaluate the optimal level of policies that would ensure maximum social-economic and sustainable development in the context of globalization
- use model solutions (weights) to guide trade-offs between various development objectives and to ensure policy coherence
- analyze the impact of changes in policies on performance efficiency via Dual LP Model.

Country's performance-policy implementation is evaluated on the basis of:

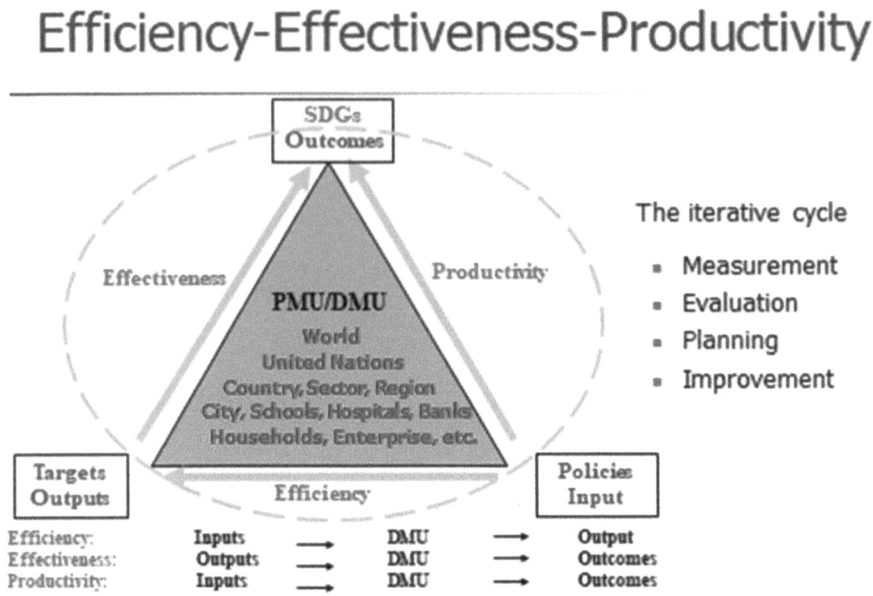

Figure 9.13 Effectiveness-efficiency-productivity cycle.

- ***Effectiveness***:
  (i) ability to state desired goals;
  (ii) ability to achieve desired goals.
- ***Efficiency***:
  (i) benefits realized;
  (ii) resources used.
- ***A multifactor productivity***:
  (i) (weighted sum of outputs)/(weighted sum of inputs).

This approach is flexible, and it is applicable to evaluating the performance and competitiveness of all sorts of DMUs: such as regions, cities, projects, and enterprises; schools; banks; etc., that are engaged in transforming multiple inputs to multiple outputs: "Frontier" PMUs (nations in our example) are able to coordinate available tools

The ultimate objective of the evaluation is to use DEA for evaluating the performance of countries and measure progress toward reducing the sustainability deficit in a globalizing world.

We propose a *three-stage DEA Model on* "A Sustainable Globalization" to evaluate:

— Global inequalities and development gaps
— Policy space and sustainable development impacts

– Development catch-up and policy developments

Step one: Measuring the global inequalities—1one input and multiple outputs

This is an indexing exercise. At this stage, a link of DEA effectiveness score with composite indicators widely used in the UN can be established. At this stage, we evaluate *the development gap*—in the form of global inequalities in countries performances! This will allow measuring the multidimensional performances of countries and providing information on policy importance given to individual variables via weights, and will yield potential improvements, identify peer groups providing a closer look at country policies.

Step two: Policy space—multiple inputs and multiple output

This is a policy analysis exercise to evaluate *the development impact*. The idea is to calculate an optimal level of policy space liberalization/protection that will enable countries to ensure maximum sustainable development impact. Target levels derived from step one of the model could be used. Using various policy variables as inputs and socioeconomic sustainable development performance variables as outputs, development impact of policies can then be assessed.

Step three: Development catch-up

This stage is already present in *policy developments* for creating opportunities over time: After we know the development gap from phase one and the optimal levels of policies, the next question to be addressed shall be: how will this outcome change if some restrictions are removed (e.g., market access) and then "what can a country do if allowed sufficient time?"

There is a wide consensus among development partners on the need for moving away from input-oriented monitoring systems and include outcome-based impact monitoring in order to shift the emphasis in assistance programs to a more systematic focus on development results, along with a more explicit linkage of development priorities and resources to results (Figs. 9.14 and 9.15).

As note earliest one of the features of DEA models is that it permits each PMU to be seen in the best possible light, compared to others in the comparison set. Comparing the results of both developed and developing nations, we propose a measure of the effectiveness of policy in converting the GDP of a nation into the UN sustainable development goals. The most successful nations in this respect (and they include not only rich, but also some quite poor nations) provide benchmarks against which the efforts of other less successful ones can be compared (Fig. 9.15).

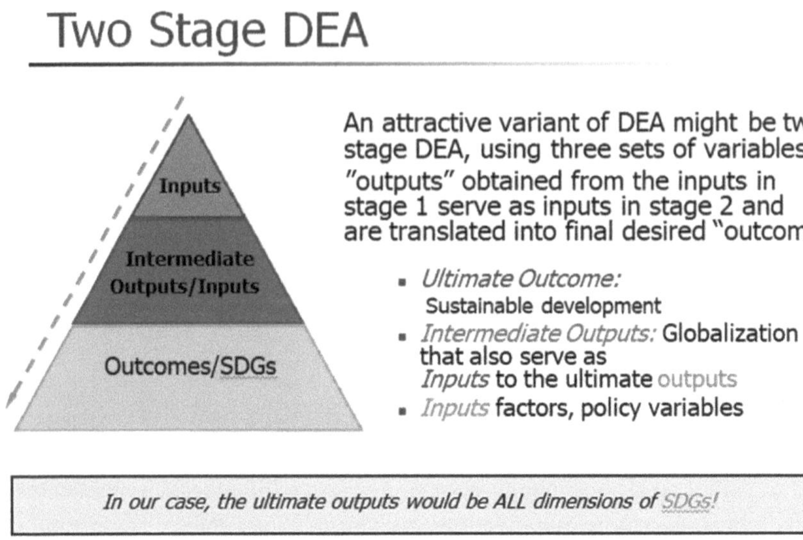

Figure 9.14 Two-stage data envelopment analysis

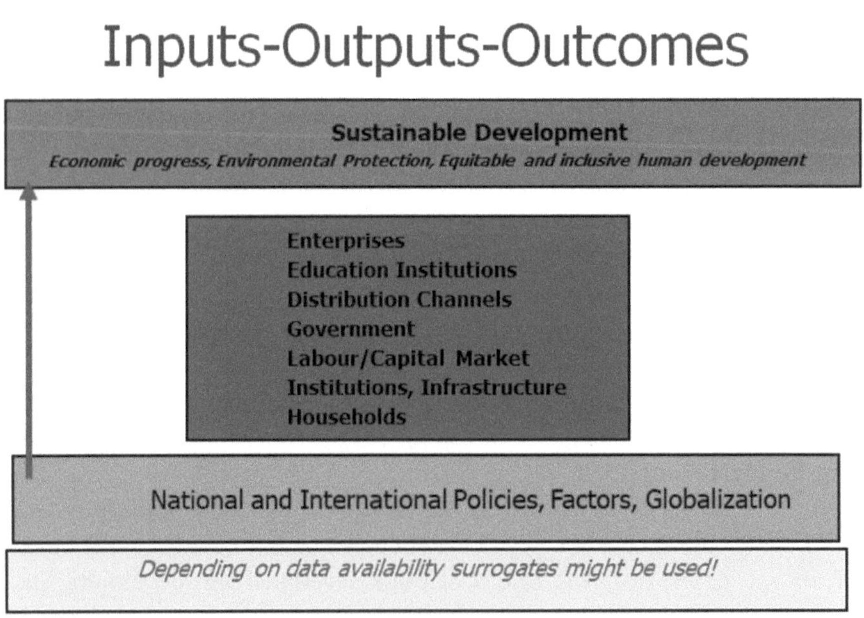

Figure 9.15 Data envelopment analysis input-output-outcome framework.

Workers' and employers' organizations have globally reinstated their strong commitment to sustainable development and to the partnership for development. Both employers' and workers' organizations have intensified their engagement and efforts in addressing the issues of environmental sustainability: the World Business Council for sustainable development, Workers Initiative for a Lasting Legacy, recognized the need for *strengthening the link between poverty reduction, environmental protection* and *decent work* and the importance of *enhancing the dialog between labor and management*, consultation and negotiations in the workplace *on sustainable development*. Major developments include: UN Global Compact, the IFC environmental and social policies, Ecuador Principles of project appraisal. Commitments to sustainable development are being encored in the strategic plans of business: examples "Lafarge-Excellence 2008"[10], Trio Tino "The way we work"[11].

Attaining sustainable development can be considered as a Social Movement entailing Social Dialog to Elicit Consensus and Policy Coherence. A consensus has emerged that achieving sustainable development calls for *synergies and coherence* of three pillars to be achieved based on principles of (1) democracy, (2) solidarity, and (3) gender equality. The Johannesburg Declaration created a *collective responsibility* to advance and strengthen interdependent, economic development, social development, and environmental protection—at local, national, regional, and global levels.[12]

Caveats: Defining homogeneous institutions and social-economic policies in terms of inputs and outputs to be used for in any modeling framework, including the DEA analysis, is a rather difficult task in view of number of local and cultural particularities arising from the existing differences in structural and institutional regimes. This considerably hinders the straightforward interpretation of model result and simple transfer of best practices making their analysis more tentative/indicative.

[10] http://www.globeinvestor.com/servlet/WireFeedRedirect?cf=GlobeInvestor/config&vg=BigAdVariableGenerator&date=20060622&archive=cnw&slug=C8429.

[11] Labor and the Environment, Toward an ILO program on the environment, Dr Francis Vorhies.

[12] The Johannesburg Declaration on Sustainable Development, September 4, 2002.

CHAPTER TEN

Toward a new social contract

The response to the pandemic, and to the widespread discontent that preceded it, must be based on a new social contract and a new global deal that create equal opportunities for all and respect the rights and freedoms of all. A new social contract within societies will enable young people to live in dignity; will ensure women have the same prospects and opportunities as men; and will protect the sick, the vulnerable, and minorities of all kinds .....The new social contract, between governments, people, civil society, business, and more, must integrate employment, sustainable development, and social protection, based on equal rights and opportunities for all .... To close those gaps, and to make the new social contract possible, we need a new global deal to ensure that power, wealth, and opportunities are shared more broadly and fairly at the international level.

(António Guterres, *United Nations Secretary-General, 2020*)[1]

10.1 Policy failures and the quest for shared prosperity

The world is in the midst of an extraordinary moment, all nations keep enormously high expectations of a more pragmatic, active global governance calling on shared interest in a better society beyond as well as within national boundaries, while absorbing the grim new reality of the sanitary crisis that has escalated into a global economic disaster, threatening the well-being of people everywhere and, sadly, undoing the recent gains in the fight against extreme poverty so many people suffer in emerging-market and low-income economies.

Sustainable development in the Covid crisis hit globalized world is, indeed, the most significant collective challenge facing humanity today and will be at the center of international research and political agenda for

[1] Secretary-General António Guterres, the 18th Nelson Mandela Annual Lecture delivered virtually on Nelson Mandela International Day (18 July 2020). The theme of the lecture is "Tackling the Inequality Pandemic: A New Social Contract for a New Era."

Measuring Sustainable Development Goals Performance
ISBN: 978-0-323-90268-7
https://doi.org/10.1016/B978-0-323-90268-7.00019-0

the decades to come. During the High-level Political Forum on Sustainable Development, convened under the auspices of the General Assembly, entitled "Accelerating the implementation of the 2030 Agenda for Sustainable Development," on 24 September 2019, *António Guterres, United Nations Secretary-General stated*: "The 2030 Agenda and the Sustainable Development Goals are our collective response to building a fair globalization."

There is a universal consensus, as supported by the empirical results presented in earlier chapters of our book, that the current model of globalization is evidently yielding uneven development results. Not all nations succeed in reaping the fruits of globalization and international expansion, a cutting edge technology. There is an absolute need for making globalization serve national interest for all countries and to create "shared prosperity"—the only kind of prosperity that can truly be sustainable, so much desired by Joseph Stiglitz.[2]

For quite some time, the key world leaders, and development practitioners, have been inspired by and preoccupied with the powerful ideas of Adam Smith manifested in *The Wealth of Nations*, assuming that government regulations were not necessary to make an economic system function in an orderly way in the interest of everyone. Only recently, economists begin to admit that the working of the price mechanism is costly, and that transaction costs strike the wheels of wealth detrimental to the well-being and the quality of life.

It was a long way for economists to grasp the message introduced by Ronald Coase in 1937, on the importance of the working of the economic system of the institutional structure of production. Coase referred to Sam Peltzman who had spoken laud about *the widespread application of theory without any empirical basis* and two Nobel Prize Laureates, Bengt Holmstrom and Jean Tirole, who in the Working Paper, "The Theory of the Firm" (Holmstrom and Tirole, 1989) published in the Handbook of Industrial Organization, identified three outstanding problems that needed attention:

1. *to develop and apply techniques that deal with nonstandard problems such as incomplete contracts, bounded rationality, and multilateral contracting.*
2. *to integrate observations from neighboring fields such as sociology and psychology—in a consistent (not* ad hoc*) way into the theoretical apparatus.*
3. *to increase the evidence/theory ratio, which is currently very low in this field.*

[2] https://www.aegon.com/newsroom/news/2019/stiglitz-making-the-case-for-shared-prosperity/.

In 2002, Coase declared that not only he thinks economics *will* change, but "he[I] think[s] it ought to change." He additionally stated that against the background of extraordinary advancements in the natural sciences such as physics, chemistry, or biology for economics:

> *there is a need for an empirical work which actually changes the way we look at the problem. What is wrong is the failure to look at the system as the object of study.*[3]

Coase acknowledged that the subject has not changed in essentials and agreed with Thomas Kuhn who in his seminal book, "The Structure of Scientific Revolutions" (Khun, 1962) provided the grounds for a paradigm shift, pointing out that *scientists aren't receptive to proposals for a change in their subject unless they are dissatisfied with the old views.*

As rightly stated by Douglass North, during his Nobel Memorial Lecture, on 9 December 1993, we shall express no mystery why the field of development has failed to develop during the five decades since the end of the Second World War. He argued that *neoclassical theory is simply an inappropriate tool to analyze and prescribe policies that will induce development* as it is concerned with the operation of markets, not with how markets develop and is grounded on two erroneous assumptions: one that institutions do not matter and two that time does not matter in the analysis of economic performance. Consequently without a clear understanding of how economies develop one cannot prescribe policies.[4]

The fact that carbon dioxide emissions heat and increase the temperature of the earth resulting in a greenhouse effect is not new. They were known to scientists for over 100 years. While favorable for earth and habitat in acceptable levels, excessive warming caused by human-generated carbon dioxide from fossil fuel burning and deforestation are increasing the atmospheric concentration of greenhouses gases, causing global climate change. During past decades, other greenhouse gases emitted by human activity have built up exponentially in the atmosphere: methane, nitrous oxide, and the same chlorofluorocarbons that resulted in a climate change leading to global warming threatening the ozone layer.

On 11 November 2013, during the UN Climate Change Conference, Mr Yeb Sano, the Philippines' lead negotiator at the UN Climate Change Convention, whose home town was devastated by super typhoon Haiyan in

[3] http://coase.org/coaseremarks2002.htm.

[4] http://www.nobelprize.org/nobel_prizes/economic-sciences/laureates/1993/north-lecture.html.

his opening address encouraged to *[we must]* "stop calling events like these as natural disasters." He more specifically underscored that:

- it is not natural when people continue to struggle to eradicate poverty and pursue development and get battered by the onslaught of a monster storm.
- disasters are never natural. They are the intersection of factors other than physical. They are the accumulation of the constant breach of economic, social, and environmental thresholds.
- disasters are a result of inequity and the poorest people of the world are at greatest risk because of their vulnerability and decades of maldevelopment.[5]

The tragic typhoon that took about 2500 human lives is another sad proof, but a strong manifestation, that shakes the long held optimistic trust in autoregulation and the persistent bias implicit it the notion of self-steering complex systems. This "natural" disaster that affected more than 11 million people and displaced some 673,000,[6] also reaffirms the importance of an improved early warning system that could have saved lives and livelihood of the victims.

On 27 November 2013, Secretary-General Ban Ki-moon, during his opening remarks at press conference with Dr. Jim Yong Kim, President of the World Bank Group following Sustainable Energy for all High-Level Advisory Board Meeting at UN Headquarters, alarmed that[7]:

> *The global thermostat is rising, threatening development goals and economies small and large. It is clear that we need a transformation in how we produce, use and share energy.*

Thermostat was a prime explanatory concept in early cybernetics. Introducing the notion of essential variables of the system, its role was to keep the preselected set of essential variable at the desired level: whereby the regulator sets a specific reference value and the thermostat "senses" the actual level by means of a thermometer and has the ability to compare it to the set value. Depending on deviation form that reference, it activates adjusting behavior. Consequently the statement of the United Nations Secretary General, who represents the interests of all "[we] the people" in the global world organized as a single United Nations System, as the prime regulator of the globe, in his

[5] https://www.unisdr.org/archive/35439.
[6] http://www.bbc.co.uk/news/world-asia-24910918.
[7] http://www.un.org/apps/news/infocus/sgspeeches/.

unique individual capacity an invitation to look at the challenges of sustainable development through the lenses of a distinct interdisciplinary science of cybernetics and systems sciences.

In 1954 in "Design for a brain," Ross Ashby defined the notions of variable, system, and homeostasis and proposed to define a form of behavior as adaptive, "if it maintains the essential variables within physiological limits" where all the quantities used in physics, chemistry, biology, physiology, and objective psychology, are variables in the defined sense. He proposed as the first step to record the behaviors of the machine's individual parts by identifying any number of suitable variables—a measurable quantity which at every instant has a definite numerical value. So the experimenter is observing not the parts of the real "machine" directly but the dials on which the variables are displayed (Ashby and Stein, 1954; Ashby, 1962).

Gregory Bateson in *Steps to an Ecology of Mind*, in 1972 defined the notion of the unit of survival expanding it from the organism to organism-in-its-environment, as:

> *the organism which destroys its environment destroys itself. The unit of survival is a flexible organism-in-its-environment.*
>
> ***(Bateson, 1972)***

Against this background, Ban's alert connects perfectly well, with what Ross Ashby submitted 65 years ago, in his seminal book *An Introduction to Cybernetics*, when presenting the challenges of regulating very large systems for most of the systems are complex and composed of almost uncountable many parts. But, as rightly acknowledged by Ashby, largeness in itself is not the source of the real difficulty: *what is usually the main cause of difficulty is the variety in the disturbances that must be regulated against* (Ashby, 1964).

In 1992, 20 years after Donella Meadows, Dennis Meadows, and Jorgen Randers outlined the prospects for growth in the human population and the global economy during the coming century, in their bestseller *The Limites to Growth*, in their next work entitled *Beyond the Limits: Confronting Global Collapse, Envisioning a Sustainable Future,* among other alerts they submitted, that:

> *on a warmer earth the ocean will expand and sea levels will rise. If the warming is sufficient to melt polar ice in large quantities, sea levels will rise significantly.*
>
> ***(Meadows et al., 1992)***

Today, we reached a stage, where the current structure of human socio-economic system has well surpassed its limits and is pregnant with a serious risk for collapse. This imminent danger justifies an urgent need for a paradigm shift to avoid further casualties and substantial structural changes at various levels of governance global, national, and local systems.

Already in 1950 Norbert Wiener, in his groundbreaking work, "The Human Use of Human Beings: Cybernetics and Society" "We have modified our environment so radically that **we must now modify ourselves in order to exist in this new environment.** We can no longer live in the old one. Progress imposes not only new possibilities for the future but new restrictions ... " (Wiener, 1950).

Already in 1964 Martin Luther King avowed in his Nobel Memorial Lecture, that:

> *the time has come for an all-out world war against poverty. The rich nations must use their vast resources of wealth to develop the underdeveloped, school the unschooled, and feed the unfed. Ultimately a great nation is a compassionate nation. No individual or nation can be great if it does not have a concern for "the least of these". Deeply etched in the fiber of our religious tradition is the conviction that men are made in the image of God and that they are souls of infinite metaphysical value, the heirs of a legacy of dignity and worth. If we feel this as a profound moral fact, we cannot be content to see men hungry, to see men victimized with starvation and ill health when we have the means to help them. The wealthy nations must go all out to bridge the gulf between the rich minority and the poor majority.*[8]

Successful implementation of Agenda 2030 explicitly requires integration of multiple dimensions of sustainable development into a single framework and there is a unanimity agreement on a need of transdisciplinary analysis embracing the complex system consisted of multiple dimensional phenomena such as globalization, poverty, and other constituents of sustainable development. The achievement of SDGs requires democratic solutions that are often not only multilateral but also multipolar and involve new or re-emerging sovereign players, and other stakeholders in the face of matured civil society representatives.

The documented failure to integrate the multiple social, economic, and environmental dimensions within a single framework (see United Nations, General Assembly, 2012; United Nations, 2015; 2015a,b,c; United Nations, 2013b) has led to a widening of inequality, both at global and national levels, pervasive social tension, alarming pace of climate change, and environmental

[8] Martin Luther King Jr. Nobel Lecture. NobelPrize.org. "Nobel Prize Outreach AB 2021." (20 Oct 2021). https://www.nobelprize.org/prizes/peace/1964/king/lecture/.

degradation (United Nations University Institute of Advanced Studies (UNU/IAS) 2002; United Nations Commission on Sustainable Development (UNCSD) Secretariat, 2011a, b). The major international declarations have now concluded that addressing this disconnect calls for a renewed role of the State in providing strategic vision and innovative policy responses toward *green growth-stimulating people-centered development strategies.*

Integrated analysis and evaluation of positive and negative impact of policy scenarios on the three pillars of sustainable development within one analytical framework involves a multiple-input and multiple-output framework. Due to the noncommensurate nature of policy and performance variables and the absence of market prices reflecting social and environmental costs and benefits, traditional analytical tools fail to capture the multidimensional character of the analysis (see United Nations Research Institute for Social Development (UNRISD) (2011)).

To better understand the analytical challenges decision makers are facing when addressing the contemporary globalization, one would wish to look at the developments in mathematics. Henri Poincaré, French mathematician, theoretical physicist, and a philosopher made a series of profound innovations and formulated what has become to be known as *the n-body problem*, and where n is any number of more than two. In 1887, in honor of his 60th birthday, Oscar II, King of Sweden sponsored a mathematical competition for a resolution of the question of how stable is the solar system. In 1888 he received the prize from the King for a question on the stability of the solar system formulated by Karl Weierstrass, renowned as the three-body (or *n*-body) problem in classical mechanics. However, Poincaré did not succeed in solving the original problem but only could invent the geometric approach to the n-body problem and *showed that in general, the stability of such systems cannot be demonstrated.* He found that the evolution of such a system is often chaotic and a small perturbation in the initial state and a slight change in one body's initial position might lead to a radically different later state.

In a highly integrated globalized economy, a very large number of entities are interacting hence the problems of stability of the system is even more difficult than n-body simulation and for that reason if n is more than two, no stability or equilibrium can be guaranteed without targeted regulation. Consequently, beyond the two-country two-goods modeling days, that dominated the analytical thinking among the researchers in most development institutions, a new model will be needed that will present the realities of the multipolar global economic space in its entirety. As mentioned earlier, Ronald Coase had criticized this failure to look at the system as the object of the study and highlighted objectively that it may well take quiet long time to do so, nonetheless to deliver the *Future We Want* for our future generations we should make an attempt to start.

10.2 The UN reform: from Versailles to Geneva and New York

Last decades were remarkable in efforts of the global decision makers in a search for more effective global government to tackle the challenges of globalization. But they were particular in recognizing that the current global system in not functional. A need for an increased role for the United Nations, the only supranational institution, that has a broad political legitimacy and an adequate representation of both middle-income countries and the least-developed countries to ensure the necessary and just regulations on key and complex issues has emerged.

There is a wide consensus that economic, environmental, and information globalization outpaced regulatory institutional and political globalization. With a remarkable progress in coordination of operations at a county level, the current model of international institutional setting and decision-making mechanisms at the global level still remain fragmented. Based on their constitutional objectives the ILO is concerned about social issues, the UNEP handles environmental domain, the IMF, and the World Bank engaged in financial and economic dimensions of development, etc. There is an open acceptance that the United Nations Economic and Social Council (ECOSOC) in its present form is unlikely to be able to accomplish the contemporary tasks of globalized world. The current structure of the governing bodies of the United Nations system organizations does not ensure a democratic and transparent organization for a world government in a multihierarchical integrated global world.

Already in 1987 Tinbergen in his article entitled *Revitalizing the United Nations System* (Tinbergen, 1987) published and with his special contribution to the 1990 World Development Report, submitted that "mankind's problems can no longer be solved by national governments. What is needed is a World Government." He suggested strengthening existing UN system, changing their nature from advisory to administrative character.

Tinbergen noted that some essential institutions crucial for global income redistribution and financial stability are still missing and particularly highlighted that as the world economy is becoming increasingly integrated so the global income redistribution shall become. Consequently he identified a need for creation of a World Treasury that through various systems of automatic taxation will connect resource and direct these revenues to cover urgent needs of individual countries where needed.

He also suggested that the existing IMF be converted to a World Central Bank, which among other things, would deal with monetary, banking, and stock exchange policies. Tinbergen identified redistribution and inequality to be a major challenge that will occupy global political agenda for the coming century and stressed that the more mature the societies are, the higher the proportion of their GDP distributed through social security benefits is, as unlike devèloping countries, the developed societies are accustomed to ruling the people and called for a convergence in this direction.

Recent financial and sanitary crises underscored the importance of social, economic, and political security and other form of transboundary risks on global sustainable development highlighted the need for policy coherence. They also stressed the need for those issues to be dealt collectively. The United Nations Millennium Declaration resolved "to ensure greater policy coherence and better cooperation between the United Nations, its agencies, the Bretton Woods Institutions and the World Trade Organization, as well as other multilateral bodies, with a view to achieving a fully coordinated approach to the problems of peace and development."

In 1991 UN General Assembly adopted a resolution entitled "Restructuring and revitalization of the United Nations in the economic, social and related fields" (United Nation, 1991) and initiated a large scale reform of the system. The main target of the on-going United Nations reform is to deepen coherence and enhance the effectiveness and impact of UN development operations, intensify efforts in pursuit of the internationally agreed development goals. It aims to enhance the system's efficiency and coherence and make the system more results-oriented at the country level. Within the new context, the UN operational activities should orientate on implementing national comprehensive development strategies that could achieve sustainable economic growth and reduce poverty.

From a chronological facet, it is noteworthy to reiterate the two historic events of the past century as rightly identified by Gregory Bateson in *Step to an Ecology of Mind* which have equally influenced my own perceptions of shared responsibility.

1. *The Treaty of Versailles* establishing the League of Nations in 1919, the forerunner of the United Nations, "to promote international cooperation and to achieve peace and security" is definitely one of the major

historical events of the past century that also the founded International Labour Organization with an explicit mandate of economic regulation to share the societal social costs and democratic supervision of the capitalist system through the means of *social dialog as an alternative path to socialist revolution* of 1917.

2. *The Cybernetic breakthrough* that provides rigorous scientific understanding of these very mysterious organized systems, offered new ways to think systematically and strategically, solve problems, and identify potential trouble spots in systems before disaster strikes and contains potential of achieving a new and perhaps more human outlook, a means of changing our philosophy of control and communication in wider perspective.

Since its inception, the United Nations General Assembly is the main deliberative, policymaking, and representative organ of the United Nations. Comprising all 193 Members States, it provides a unique forum for multilateral discussion of the full spectrum of international issues covered by the Charter signed at San Francisco on 26 June 1945. The UN has always been perceived to be the only democratic international forum that enables all nations to enjoy the right to express freely and engage on the basis of sovereign equality.

There is a growing consensus that as the world's only legally constituted and globally inclusive intergovernmental body with a clear mandate the UN General Assembly has a special and unique role to play in our global deliberations and an explicit responsibility for effectively coordinating the work of the specialized agencies at the intergovernmental level along with the ECOSOC that was established in as a standing committee of administrative officers of the specialized agencies for improved coordination.

Distinct from ECOSOC, the UN System Chief Executives Board for Coordination (CEB[9]) originated in 1946 as the main instrument for supporting and reinforcing the coordinating role of the United Nations intergovernmental bodies on social, economic, and related matters to strengthen synergies, identify and reduce duplication and gaps, and promote coherence and coordination among the organizations of the UN system.

It is the longest-standing and highest-level coordination forum of the United Nations system. CEB is chaired by the UN Secretary-General and comprises 29 Executive Heads of the United Nations and its Funds and Programmes, the Specialized Agencies, including the Bretton Woods Institutions

[9] http://unsceb.org/content/ceb.

(The World Bank and IMF), the United Nations; 15 Specialized Agencies established by intergovernmental agreements; World Trade Organization and the International Atomic Energy Agency; and 11 Funds and Programmes created by the United Nations General Assembly and UN Women.

CEB meets biannually to provide broad guidance, coordination and strategic direction for the system as a whole in the areas under the responsibility of executive heads and carries out its role through three high-level committees:

1. High-Level Committee on Programmes (HLCP) that promotes policy coherence and system-wide cooperation, coordination, and knowledge sharing in strategic programme areas,
2. High-Level Committee on Management (HLCM) identifies and analyzes administrative management reforms with the aim of improving efficiency and simplifying business practices, and
3. The United Nations Development Group (UNDG) responsible for coordinating operational activities at the country level.

Focus is placed on interagency priorities and initiatives while ensuring that the independent mandates of organizations are maintained.

In 1948, the UN Secretary-General established the Administrative Committee on Coordination ACC. Under the chairmanship of the Secretary General, the CEB provides coordination and strategic direction for the system as a whole focusing on inter-agency priorities and initiatives, while ensuring that the independent mandates of organizations are maintained. The CEB meets twice a year and in addition to session reports, an Annual Overview Report summarizes the collective impact of member organizations.

A Report of the Commission of Experts of the President of the United Nations General Assembly on Reforms of the International Monetary and Financial System published in September 21, 2009 stressed:

> *Reform of the international system must have, as its goal, the improved functioning of the world's economic system in support of the global good.* ***This entails simultaneously pursuing long term objectives,*** *such as sustainable and equitable growth, creation of employment in accordance with the "decent work" concept, responsible use of natural resources, reduction of greenhouse gas emissions, and more immediate concerns, including addressing the challenges posed by the food and financial crises and global poverty.*
>
> *It has made clear that globalization of trade and finance* ***calls for enhanced global cooperation and global regulation.***
>
> ***(United Nations, 2009)***

The report noted that the crises, as well as the growing divide between poor and rich within and between countries and the risk of systemic climate change, are all interconnected global challenges that threaten to unravel the fragile state of globalization and called upon the UN reform to take appropriate measures for acceleration the progress toward these overarching priorities and deliver on the longstanding commitments to achievement of the internationally agreed development goals and protecting the world against the threat of climate change.

There is a universal acceptance that attaining sustainable development calls for a convergence between the three pillars of economic development, social equity, and environmental protection. There is a need to develop a new decision making framework that will be capable to deliver on widespread concerns of equity, poverty alleviation, and integrating economic, environmental, and social dimensions of development within a single agenda.

The newly emerged Global Partnership for Effective Development Cooperation calls for a more inclusive, efficient, and effective global governance. With recognition of a central role for the UN System in the coordination and monitoring of the implementation of global sustainable development goals (SDGs), a new institutional framework needs to be developed to adequately respond to the current needs of a contemporary globalized world.

The Rio +20 outcome documents the *Future We Want* resolved to establish an inclusive and transparent intergovernmental process on SDGs that is open to all stakeholders with a view to developing global SDGs to be agreed by the General Assembly of the UN. An intergovernmental Open Working Group was established to this end. The outcome document specifies that the process leading to the SDGs needs to be coordinated and coherent with the processes considering the post 2015 development agenda and that initial input to the work of the Open Working Group will be provided by the UN Secretary General in consultation with national governments.

But, we have been beyond the acceptable limits for our one and only planet for quiet sometimes now, and unless we learn the lessons from the precybernetic and Washington Consensus era, and continue to operate on the premises of the *logical* reductio ad absurdum and look into regulation of the chaotic processes, prevailing in the real world, instead of ensuring the aspirations stipulated in the *Future We Want*, these outdated approaches will drive the current highly integrated world into a massive ecological and humanitarian catastrophe.

Jointly with his scientific companion Arturo Rosenblueth, Nobert Wiener made an attempt to embrace the entire field of communication, control, and statistical mechanics by a single term. In 1948, in his groundbreaking book, *Cybernetics: Or Control and Communication in the Animal and the Machine*, he developed a new branch of applied science—the science of information feedback systems, setting off a scientific and technological revolution. Wiener defined cybernetics as the theory of messages. It encompasses four distinguishing components: *variety* that involves all of the possible combinations of input and output in a given system, *circularity* that refers to self-regulation based on feedback, *process* demonstrating how the system works, and *observation* that involves decision making within the system. When defining the term "feed-back," he gives the classic picture of the microstructure of action:

> ***When we desire a motion to follow a given pattern, the difference between this pattern and the actually performed motion is used as a new input*** *to cause the part regulated to move in such a way as to bring its motion closer to that given by the pattern.*
>
> ***(Weiner, 1961)***

Wiener further in *The Human Use of Human Beings: Cybernetics and Society* submitted:

> *This control of a machine on the basis of its* ***actual performance*** *rather than* ***expected performance*** *in known as feedback.*
>
> *The information is then turned into a new form available for the further stages of performance ... this performance is made to be effective on the outer world. In both of them, their* ***performed action*** *on the outer world, and not merely their* ***intended action,*** *is reported back to the central regulatory apparatus.*
>
> ***(Wiener, 1989)***

The Secretary General's High Level Panel on the Post-2015 Development Agenda reiterates the vital importance of building data systems to provide timely, disaggregated indicators to measure progress, in all countries, and at all levels (local, subnational, and national).

The availability of statistical information has considerably improved during the implementation of the millennium development goals (MDGs). Learning from data analysis—and adapting actions based on what we learn from evaluation—is one of the best ways to ensure that development goals are reached. The revolution in information and communication technologies over the last decade provides an opportunity to strengthen data collection and statistical analysis for accountability and decision-making purposes.

Mobile technology and other innovative initiatives and advances have also been used and to enable collection of real-time data.

Better data and statistics will be a considerable help to governments to track progress and make sure their decisions are evidence based. They can also strengthen accountability not just for national governments but also international agencies, civil society organizations (CSOs), and the private sector should be involved. A true data revolution would draw on existing and new sources of data and will fully integrate statistics into decision making, promote open access to, and use of, data, and ensure increased support for statistical systems.

Toward this end, the Panel's report entitled "A New Global Partnership: Eradicate Poverty and Transform Economies through Sustainable Development," released in 2013 recommended establishing a Global Partnership on Development Data that will bring together diverse but interested stakeholders—government statistical offices, international organizations, CSOs, foundations, and the private sector (United Nations, 2013).

This partnership would, as a first step, develop a global strategy to fill critical information gaps, expand data accessibility, and galvanize international efforts to ensure a baseline for informed decision making for the post-2015 targets is in place by January 2016. Investments in data gathering will be needed to ensure data for indicators that are not yet available. Technical experts meetings are foreseen to discuss and refine the models and methods for qualitative indicators that are not yet agreed or are unclear.

Data envelopment analysis (DEA) co-founded by Bill Cooper provides a strong foundation for integrating discipline-based knowledge into a single analytical framework that can become an analytical tool for developing more specific, operationalizable global governance models. Within the formal structures of global governance, negotiations and consensus building processes DEA-based models will yield optimal solution for the global world, going well beyond global economy to cover the challenges facing the environment, global society, etc.

While cybernetics does not directly discuss the role of the observer, the distinguishing feature of DEA is that it provides a coherent cycle of data analysis performance enhancement quiet harmonious with information feedback behavioral change.

This interesting feature of envelopment analysis originated in 1978 explains its wide-scale practical application to enhance performance of decision-making unit (DMU) of totally different nature. The DEA-based performance evaluations have culminated in the expanded version of its application to measuring policy effectiveness of countries as presented in the Part 1 of the book.

Moreover, the DEA model solutions, containing the optimal levels of variables, slakes, efficiency losses, identification of peers, etc., connect perfectly well with the cybernetics concept of learning as an indirect method of regulation, and the adoption of a pattern of behavior that is compatible with survival. This very exceptional advantage of DEA: to account for improvements and technological chance, was long disregarded in earlier analytical frameworks and economic theories.

In fact the establishment of General Agreement on Tariffs and Trade (GATT)/World Trade Organization (WTO) and its continuous deliberation toward creation of level playing field is a manifestation of global political will to reduce the differences among the territories to be reflected on the global map. Consequently when looking with cybernetic lenses at the standard setting and normative deliberations of the UN System, they should be viewed as manifestation of enabling environment for reducing the variety in the disturbances in the system of the globalized interconnected world, hence contributing to a decrease in entropy that in turn will increase the predictability in the global trading system and serve as a guarantee for global stability. From its inception the GATT/WTO was built on the practice of consensus decision-making, the major challenge remains, however, on how to prevent the consensus building process from unjustified exercise of veto without giving a reasoned explanation for their opposition.

The proposed DEA-based "Geneva Consensus" framework provides a solution to these problems. It is grounded in the principle of collaborative consensus building among already WTO 164 member countries[10]. Based on cybernetic feedback processes and informed evidence-based negotiations, mutual interests could be accommodated. The proposed governance framework will enable to effectively use the institutional capacity of the WTO and other international organizations and ensure that neither issues opposed by many are pushed nor progress desired by many blocked.

Enhanced data availability will create the information basis for development of divers DEA-based global cybernetic models, thus providing an analytical underpinning for global governance.

[10] https://www.wto.org/english/thewto_e/whatis_e/tif_e/org6_e.htm.

The advent of information and communication technologies, the methodological advancement of statistical development and analysis, and most importantly the "data revolution" called by the report of the UN Secretary General's High Level Panel will create legal, political, and methodological grounds for such a change (United Nations, 2013).

10.3 Beyond a broken social contract

The 17 SDGs that are at the heart of the *2030 Agenda for Sustainable Development* have now been perceived as the world's best plan to build a better world for people and our planet by 2030. Through partnerships and peace, goals interconnect and apply to all countries.

On 24 and 25 September 2019, Heads of State and Government gathered at the United Nations Headquarters in New York to follow up and comprehensively review progress in the implementation of the 2030 Agenda for Sustainable Development and the 17 SDGs. The event was the first UN summit on the SDGs since the adoption of the 2030 Agenda in September 2015.

There had been a firm consensus that governments, alone, cannot deliver it and they are reaching to forge new partnerships with civil society, businesses, and the scientific and technological community. There was recognition, that for such efforts to be effective it is imperative that all countries pursue the SDGs as systems, fully recognizing the interconnections across sectors countries and levels, from global to local. Positive synergies need to be harnessed and difficult trade-offs need to be addressed more quickly and effectively.

During the Forum, the United Nations members have declared that 2020 marks the start of a Decade of Action to fulfill the objectives of the 2030 Agenda and its SDGs. Goal 8—on decent work and sustainable growth—is at the core of this agenda. It is instrumental to reverse global injustice and realize a new social contract with decent work for all.

The summary documents of the High Level Political Forum, convened under the auspices of the General Assembly SDG Summit, held on 24–25 September 2019, reiterate the necessity of a systemic and all-inclusive approach:

> *Focusing on one goal or target at a time and working in silos is therefore unlikely to result in overall success. Transforming to a socially, economically and ecologically sustainable society requires strengthening the focus toward all the SDGs and ensuring cooperation among the different measures or levers of action.*

> *The widely accepted ceteris paribus condition, which analyses the behavior of each system independently, cannot be considered helpful in the Agenda 2030 context. More holistic approaches are required as the contextual evaluation of several systems together is fundamental.*
>
> ***(United Nations, 2019)***

The 2030 Agenda has now been qualified by the Forum as a plan of action for people, planet, prosperity, peace, and partnership—a plan to free humanity from the tyranny of poverty and heal and secure our planet for future generations.

But, there is a widespread critic that the social contract is broken! World Bank (2015a). In July 2020, United Nations Secretary General António Guterres called for a new global deal as "the only way that we will meet the goals of the 2030 Agenda for Sustainable Development, the Paris Agreement, and the Addis Ababa Action Agenda—agreements that address precisely the failures that are being exposed and exploited by the pandemic."

Since the start of the COVID-19 pandemic, the call for a "new social contract" has also intensified among all major development organizations, ranging from Save the Children to the World Economic Forum. Recognizing that present ways of developing policy and executing financing are broadly not working Guterres called for a new, long-term social contract between governments and societies to address rising levels of global inequality.

COVID-19 is a human tragedy, indeed. But as the UN Secretary General views: it has also created a generational opportunity to build a more equal and sustainable world, based on two central ideas: *a new social contract and a new global deal.* A new social contract will link together governments, their people, CSOs, businesses, and others in common causes. He expressed a hope that the new global deal and the new social contract will put the world back on track to realize the promise of the Paris Agreement on Climate Change and achieve the SDGs—the globally agreed vision of peace and prosperity to be achieved by 2030.

The widespread unemployment, persistent poverty, vulnerability and misery, are a distinct manfestation of the absence of effective social protection system, and make it clear that the old social contract is broken. New sets of policies will be required toward more enabling environment for the operation of the private sector. Among these changes, the most important will be to promote competition in domestic industries, so that young firms can grow and create productive and decent jobs.

The second identified area where the social contract appears to have broken down is in delivering quality public services. It is clear that the old social contract, although it delivered results in the past, is not suited to the needs of the current generation of citizens.

World Bank estimates the exploitation of workers in global supply chains with more than 60% of global trade dependent on precarious contracts. This is simply inequality by design with an economic model based on low wages, insecure, and often unsafe work. 60% of the world's workers are now working in informal work, including new platform businesses, with no rights, no minimum wages, no social protection, and no rule of law World Bank (2015a).

With our own empirical analysis, we highlighted the fundamental role of decent work in poverty reduction and sustainable development (Thore and Tarverdyan, 2009). And made an attempt to identify policies conducive to decent work, it is not surprising, that workers want a New Social Contract with SDG 8 at its core and decent work for all! Achievement of SDG 8 on decent work is clearly of paramount importance for the global labor movement. *Sharan Burrow, general secretary, International Trade Union Confederation* presents the interests of the workers who "want a New Social Contract that delivers decent work for all."

With an understanding, that the ILO Centenary Declaration, ILO (2019) including the labor protection floor for all workers, would go a long way to rebuilding the social contract, workers express a need and commitment to hold governments accountable to their commitments in delivering the SDGs and track progress of own governments policy interventions. Unions reaffirmed, they stand ready to fight to ensure that full employment and decent work remain the foundations of any economy and a new social contract is a vital start if we are to rebuild trust in a fractured world severely hit by Covid crisis.

According to the estimations of the International Trade Union Confederation in global supply chains, 94% of the global workforce is a hidden workforce where the obscurity of business contracts facilitates exploitation and oppression https://www.ituc-csi.org/new-ituc-report-exposes-hidden. They expressed great dissatisfaction that the global economic model has failed working people. The power and greed of huge global corporations have captured governments, which are acting against the rights and security of their own workers.

The ITUC submits that even though the world is three times richer than 20 years ago, yet 70% of people are denied universal social protection, 84% of people say the minimum wage is not enough to live on and 81% of countries have allowed violations of the right to collectively bargain. As ITUC claims it, *this is inequality by design.* http://www.sharing4good.org/article/new-social-contract-recovery-and-resilience. They call for a new social contract between workers, government, and business *which should include a floor of a universal labor guarantee for all workers.* The trade unions expect all dimensions of the decent Work Agenda to be implemented in the framework of the new social contract: labor rights, jobs with decent minimum living wages, collective bargaining, universal social protection, business operations driven by due diligence and accountability, and social dialog ensuring just transition measures for climate and technology.

The social contract is as commonly defined, as incompacing the shared understanding of the roles and responsibilities of each major stakeholder in the economy and the guiding principles they use to carry them out. The new social contract must therefore be built to address the challenge of our time, and it must be equal parts of public and private sector solutions.

In the context of the Fourth Industrial Revolution, the new social contract shall begin by businesses creating more inclusive economic growth and shared value for workers and communities.

The Post Covid Great Reset imposes on us the need to forge a new social contract in which workers, businesses, and government, together with other stakeholders, share in the opportunity and the risk of the new economy. This implies reconsideration of the roles and responsibilities of each of these stakeholders and reinvention of the tools of action used to manage risk. Appropriate innovative packages of public policy and market-based solutions must be identified.

It is noteworthy that globalization and economic integration have reduced the national policy space to regulate labor conditions and imply international actions to manage changes in the world of work under a new social contract. COVID-19 has exacerbated the employment crisis and in the new circumstances, only by the means of international cooperation, solidarity, and cooperation of the new social contract can be implemented effectively.

Markus Loewe et al. in a recent study present the social contract as a tool of analysis (Loewe et al., 2021). They recall that the concept of the social contract was coined by the liberal state philosophers Grotius, 1625, Hobbes, 1951, Locke, 1689, and Rousseau (1762) and since then been extensively discussed in a large strand of literature. Most writings though are rather theoretical and normative in that they consider the social contract as something that is good or even necessary to overcome the natural state of anarchy, and to establish property rights and security or distributive justice.

They identify various levels of the contract: national contracts (Kaplan, 2015), supranational agreements, like the European Union (Rhodes and Mény, 2016), global social contracts (Birdsall, 2008; Stone, 2014). Certain authors analyze what deliverables state and society can exchange in a social contract (e.g., Avdagic et al., 2011; Bukenya, 2016; Cook and Dimitrov, 2017; Feldmann and Mazepus, 2018; Hickey and King, 2016; OECD 2009; Regeni and Auktor, 2017).

10.4 The Calculus of consent, group rationality, and Pareto optimality

James M. Buchanan is considered one of the greatest scholars of liberty of the 20th century, an eminent economist who won the Nobel Memorial Prize in Economics in 1986 as a pioneer, as well as the leading person, in the new field of research that is now generally called *Public Choice* or the new political economy, more specifically: "for his development of the contractual and constitutional bases for the theory of economic and political decision-making."[18]

He earned his prize, in large part, because of his path-breaking work in the analysis of political institutions from an economic perspective. His major contributions to public choice theory bring together insights from political science and economics to explain how public-sector actors (e.g., politicians and bureaucrats) make decisions. He showed that contrary to the conventional wisdom that public-sector actors act in the public's best interest (as "public servants"), politicians and bureaucrats tend to act in their own self-interest, just like private-sector actors (e.g., consumers and entrepreneurs). He described his theory as "politics without romance."

The Presentation Speech by Professor Ingemar Ståhl of the Royal Academy of Sciences, states, the prize committee noted: "The political process ... becomes a means of cooperation aimed at achieving reciprocal advantages. But the result of this process depends on 'rules of the game[11], i.e., the constitution in a broad sense. This, in turn, emphasizes the vital importance of the formulation of constitutional rules and the possibility of constitutional reforms. According to Buchanan, it is often futile to advise politicians or influence the outcome of specific issues. "In a given system of rules, the outcome is to a large extent determined by established political

[11] The Nobel Prize. "Press release (1986)". https://www.nobelprize.org/prizes/economic-sciences/1986/press-release/.

constellations. A relevant example is that those who would like to correct individual tariffs should concentrate instead on the fundamental rules of international trade, such as GATT."

Buchanan and Tullock's *Calculus of Consent: Logical Foundations of Constitutional Democracy*, with its great originality and significance represents a major step toward formulating a scientific theory of democracy. I will not be the first one to say that the Calculus of Consent is a vital contribution to the literature concerning the analysis of political decision-making processes with the application of economics methods. All individuals as members of households or of firms act mainly in their self-interest. Nevertheless the outcome of the whole system is usually in accordance with the "general interest." The same individuals that act mainly in their self-interest as members of households and firms and seek to increase utility or profit are also members of the political system. An important conclusion of this strain of research is that attempts to correct different economic coordination or "market failures" by political and administrative measures might instead create new and possibly equally serious "policy failures." A further consequence of this research is a social theory that not only integrates political and economic decision-making but also includes important parts of the legal rule-system.[12]

This ground-breaking economic classic widely known as the Economic Theory of Politics presents the basic principles of Public Choice theory, where the relevance of Pareto's general theory of social interaction for contemporary Public Choice analysis was elaborated. As a unique blend of economics and political science, it created significant new subfields in both the public choice school and constitutional political economy.

Analyzing the traditional political science approach to voting systems, Buchanan and Tullock demonstrated that, since there is always a tradeoff, neither the orthodox model of majority rule nor "the unanimity rule" is perfect as a standard system: while the former imposes varying amounts of both external costs and decision-making costs the latter has little or no external costs, but has considerable decision-making costs. They thus conclude that decisions with potentially high external costs should require unanimity or at least supermajority systems. The authors based their analytical approach on methodological individualism that implied that collective action is composed of individual actions and rejected any organic interpretation of the state, presuming

[12] The Nobel Prize. "The Sveriges Riksbank Prize in Economic Sciences in Memory of Alfred Nobel 1986." https://www.nobelprize.org/prizes/economic-sciences/1986/ceremony-speech/.

that: "the state is an artifact, created by men and thus subject to change and perfection." *Unanimity* has always been perceives as an agreement by all people in a given situation whereby groups may consider unanimous decisions as a sign of, e.g., social, political, or procedural agreement, solidarity, and unity. Another fundamental point raised by Buchanan and Tullock was that: "only constitutional changes, which can be shown to be in the interest of **all** interested parties can be judged as 'improvements' and therefore consider conceptual unanimity as the only legitimate decision-making rule."

The Calculus of Consent breaks with past theories of political science in its analysis of democratic decision-making processes and it approaches the basic problems of politics Buchanan and Tullock diverged from political scientists, who defined the political process as a system in which the policy decisions are viewed as a private interest versus public interest struggle, and viewed the public interest as an aggregation of private decision makers, using the technical tools developed in modern economics and game theory.

At the heart of their thinking, Buchanan and Tullock make parallels among the political institutions and the market considering the Government as a co-operative endeavor on the part of a number of people of differing tastes to increase their abilities to reach their separate objectives. Going back to the work of Arthur Bentley and Pareto, they identify the government's task to be section of set institutions that will best serve the individual ends of the citizens, in an analogy with the basic question of efficiency widely used in economics, considering the individual as he participates in the processes through which group choices are organized.

Over a century, the *Pareto Optimality criterion*, developed by Vilfredo Pareto, has been employed by welfare economists in determining whether or not a given situation is "efficient" or "optimal" and to assess whether or not a given move or change is "efficient" or "optimal."

Buchanan and Tullock submit that welfare-political-economist approach indicates that *a specific choice is Pareto optimal only if all parties reach agreement and admit that for any single decision or choice, full agreement must be possible if the action is to be justified by the Pareto rule*. They define the "welfare" of the whole group of individuals to be increased if:

(1) WIN-WIN: every individual in the group is made better off, or

(2) WIN-NO LOSE: if at least one member in the group is made better off without anyone being made worse off.

The situation is, then, defined as Pareto optimal or Pareto-efficient, if in any given situation:

NO LOSE: it is found to be impossible to make any change without making some individual in the group worse off.

The situation is defined as nonoptimal, in a given situation:

WIN-NO LOSSE: if it is found possible to make at least one individual better off by a change while making no individual in the group worse off.

The first use of the Pareto norm is, therefore, to provide a means of classifying all possible social states or situations into the Pareto optimal set and the nonoptimal set.

The second application of the Paretian construction lies in the development of a rule for classifying changes in social situations. A change is defined to be Pareto-optimal if, in the transition from one situation to another, either

(1) WIN-WIN: every individual in the group is made better off, or
(2) WIN-NO LOSE: at least one individual in the group is made better off and no one is made worse off.

According to them, the welfare or political economist may construct a hypothesis about operational propositions of specifically proposed policy changes; and advance a proposal as "presumed Pareto-optimal," however, this proposal is subject to testing and conceptual refutation that may be tested empirically. While the attainment of consensus in support of the change would lend support to the hypothesis, failure would tend to refute the hypothesis.

They acknowledged that modern welfare economics has been concerned primarily with collective action at the concrete level, and while certain attempts have been made to devise criteria for judging specific policy measures and even though this point has not been developed sufficiently: *the reaching of unanimous agreement is the only possible test for improvement in the restricted Pareto sense.*

For Buchanan and Tullock, in terms of the constructions of modern welfare economics, the whole question of the relationship between the operation of simple majority voting rules and the "efficiency" in resource usage can best be discussed within the context of the game-theory models.

10.5 Consensus as a norm

A quest of an acceptable mapping from the set of individual perspectives and the problem of finding consensual values as an aggregation problem has been at the heart of Social Choice theorists. Since the postulations by

Kenneth Arrow, they have been seeking reasonable general conditions on any such aggregation mechanism or function that would allow to identify the set of mappings worthy of consensus. Arrow's "impossibility" theorem—or "general possibility" theorem, as he called it—postulates a very basic question in the theory of collective decision-making: there are some alternatives to choose among policies, public projects, candidates in an election, distributions of income, and labor requirements among the members of a society, or just about anything else.

There is a wide range of literature on the method for estimating a group consensus (Zahedi, 1986; Goicoechea et al., 1982; Eliashberg and Winkler, 1981). Both Pareto efficiency and consent have an inherent appeal. At the heart of Paretian analysis is *WIN-WIN paradigm*, the stipulation that no one loses in those situations which are Pareto preferred; and the logic of consent carries with it the conclusion that consented-to arrangements are to everyone's perceived self-interest. As a result, there is a tendency to assume that both Pareto efficiency and consent enjoy a significant connection to justice.

In their groundbreaking work in democratic theory, by describing the individual calculus on the constitutional level, Buchanan and Tullock fuse economic methods, political theory, and the normative project of showing how democratic institutions of a particular sort can be justified contractually, and come up with an innovative techniques to solve the fundamental problem of democratic justification, suggesting a strong condition for consensus building

> *Any rule for collective choice embodying less than full consensus must impose some external costs on the individual since resources will tend to be allocated "inefficiently" because of the choice mechanism. One means of escape from what appears to be a hopeless methodological dilemma is that of introducing some rule for unanimity or full consensus at the ultimate constitutional level of decision-making.*

As a pioneer and the architect of public choice and constitutional political economy, Buchanan, contributed to many fields of study, including philosophy, political science, and public finance. He preferred the political economy framework to pure economics and was focused on people's decision-making process within the political realm. He combined both the fields of economics and political science and was inspired by the analysis of the wealth and income distribution effects on society. Divorcing much from the long and continuing debate concerning the validity of majority rule, they created a unique form of democratic contractualism that came to be known as "Constitutional Political Economy." Viewing the politics as a system of choice and subsequent-related actions, and thercafter, the political economy as a domain dealing with the understanding of collective decision making, they smoothly and fruitfully integrated Paretian economic sociology into the Public Choice framework. Their individualistic theory of the constitution, thus, considered Pareto's theory of individual and collective choice in this light and assigns a ***central role to a single decision-making rule—that of general consensus or unanimity.*** More specifically, at the ultimate stage of constitutional choice, when decisions must be made among alternative means of or gaining human activity and among rules for collective decision-making, made:

> *full consensus of unanimity among all members of the social group to be the only conceivable test of the "rightness" of the choices.*

In their own words, this postulated unanimity rule for ultimate constitutional decisions allows us to us to divorce much of our analysis from the long and continuing debate concerning the validity of *majority rule as an absolute doctrine of popular sovereignty*. Buchanan and Tullock, believe that, the formation of a "social contract" on the basis of unanimous agreement becomes possible, insofar as participation in the organization of a community, a state, is mutually advantageous to *all* parties. Moreover, the only test of the mutuality of advantage is the measure of agreement reached.

Concentrating their attention to nonoptimal conditions, and making an analogous to the Pareto rule for assessing changes in welfare economics, they submit that, a nonoptimal position suggest that there exists a means of moving to an optimal position in a Pareto optimal manner. Thereafter, applying this fully analogous reasoning to the contractarian terminology, they conclude that the definition of an existing set of political rules as reflecting consensus implies only that there exist no particular changes on which all citizens can agree. The definition of an existing set of institutions as nonoptimal in the sense that it does not reflect consensus means strictly that changes are possible on which *all* members of the group may agree.

As rightly noted by Buchanan and Tullock, individuals cannot competently choose between collective and private action in a particular area until the results of alternative choices are analyzed and given the nature of the collective action, the ultimate decision-making authority must be specified before it can be properly assessed as an alternative to private choice. To deal with methodological dilemma, they propose introducing some rule for unanimity or full consensus at the ultimate constitutional level of decision-making and criterion against which the individual person's decisions on constitutional issues may be analyzed:

> *in examining the choice calculus of the single individual, as this calculus is constrained by the knowledge that all other individuals in the group must agree before ultimate action can be taken, we are able to discuss meaningfully **"improvements"** in the rules for choice-making.*

For them, the only real measure of "improvement" that may be accomplished through change is the agreement among all individuals in the group upon the change become:

> *The only means whereby the individual can insure that the actions of others will never impose costs on him is through the strict application of the rule of unanimity for all decisions, public and private.*

Within their welfare-political-economist approach, a specific choice is Pareto optimal only if all parties reach agreement and presumes that even the most rationally constructed constitution will allow some decisions to be made that are "nonoptimal" in the Pareto sense. Taking pride in their ability to describe the individual calculus on the constitutional level, they identify conditions under which consensus can be forged or refuted:

> *The welfare or political economist may construct operational propositions about specifically proposed policy changes; he may advance a proposal as "presumed Pareto-optimal." This proposal then takes the form of a hypothesis subject to testing, subject to conceptual refutation. The test lies in the degree of support that the proposal obtains. The attainment of consensus in support of the change would lend support to the hypothesis; failure would tend to refute the hypothesis.*

Consequently, they distinguish between the empirical validations of specific propositions advanced by the political economist, to insure that a proposed change is, in fact, Pareto-optimal, for the attainment of unanimous support from the notion that the rule of unanimity should be chosen at the constitutional level as the appropriate decision-making rule for collective choices.

In "What Should Economists Do?" states (Buchanan, 1979):

> *If we reject the notion that there must exist a public or general interest apart from that of the participants, we are necessarily led to the conclusion that **only upon unanimous consent of all parties can we be absolutely sure that the total welfare of the group is improved.***

In his other book, "The constitutional way of thinking" published in *Supreme Court Economic Review*, (Buchanan, 2003) he submits:

the economists have been forced into the sometimes tortuous searches for nonexistent social welfare functions. The political theorists, since the ancient Greeks, have had no comparable disciplinary bridles. They have felt no compulsion to derive collective-political objectives, or means to reach such objectives, from individual evaluations. Individuals are not the elemental building blocks for a normative theory of political order. From the beginning of any philosophical inquiry, political theorists have commenced with the collective unit, the polis, the city, the state, and discussion has been concentrated on the discovery of that which is "good" for such an entity.

10.6 A utility function nonexistent until "discovered"

As mentioned earlier, in a 2010, our close collaboration with William Cooper, culminated in a discovery of surprising connection between maximizing a social welfare function and DEA. A joint article entitled "A utility function approach for evaluating country performances with DEA – the twin goals of decent work and a fair globalization," one of Bill's last published methodological papers, was printed in the volume Global Operations Management, ed. L.R. Hockley (Cooper, Thore and Tarverdyan, 2010).

Applying the original DEA model, using all statements being in logarithmic units, a new interpretation of the fractional programming problem of DEA emerged: maximizing the utility/cost ratio for a single unit to be ranked, subject to the condition that this ratio remains feasible for every other unit. Applying the same approach, in Part 1 of the current book, we aggregated the SDG data for a single nation into a representation of its "SDG welfare function." Using the mathematical machinery of DEA, we have evaluated the effectiveness of countries' efforts in converting *their GDP to SDG achievements,* and identified Pareto optimal and nonoptimal set of countries.

Reverting back to the Brundtland Commission report definition of *Sustainable Development*, as "development that meets the needs of the present without compromising the ability of future generations to meet their own needs" (see The World Commission on Environment and Development, *1987*), we shall recall again, that Pareto efficiency or Pareto optimality is a situation that cannot be modified so as to make any one individual or preference criterion better off without making at least one individual or preference criterion worse off. Consequently from the scientific perspective, ensuring the shred "shared prosperity" makes a case of looking for Pareto optimum solutions.

In fact all Nobel Laureates referred to above, expressed critical view on the theoretical background of economics, stressed a need for more empirically supported analysis, highlighted the central role of human development and behaviors for economic system, and acknowledge the complexities of existing systems. They, called for a paradigm shift and application of economic models and interdisciplinary and systems approaches.

As we will present it bellow, the answer to all above concerns lies at the heart of DEA-based models. The approach proposed here is an attempt to use DEA (mathematical programming) models as a nonparametric alternative to currently used econometric models and links operations research with general economics as an aid to decision-making and political consensus building. We submit that DEA first introduced by Charnes et al. (1978) comes closer to solving the n-body problem, thanks to its key property: DEA integrates n distinct DMUs into one—a single modeling framework, and hence it simultaneously solves n optimization problems, one for each of n-DMUs under consideration within the unified model. Based on the available statistical data, DEA models empirically derive optimal solutions for all DMUs involved that jointly and individually satisfy conditions of Pareto-Koopmans efficiency or effectiveness, as defined earlier, when the original observations are replaced by their logarithms.

The game theory has widely been used in economics and political science, as a computational means for formulating a strategic interaction between multiple agents where the decision of each player is based on their actual utility or outcome. It encompaces the process of modeling the strategic interaction between two or more players in a situation containing set rules and outcomes. A set of notions including players, strategies, utilities, and game rules are utilized in an attempt to mathematically model the behavior that governs a strategic situation treating conflict of interest. Within the context of Buchanan's Calculus, DMUs might be viewed as players in a game, and each DMU may choose to take a game to maximize its utility. Therefore, viewing DEA from the lense of cooperative game theory will provide an instrument to analyze whether incentives for cooperation are present or to allocate the payoff (gain or cost) of a game and will yield a fair and unique solution plan for all simultaneously competing and cooperating DMUs. This is a cooperative game approach and can be used to the multistage systems. According to cooperative game theory, or centralized control, the two stage process can be viewed as one where the stages jointly determine a set of optimal weights on the intermediate factors to maximize their efficiency scores.

10.7 Pareto-Koopmans optimality criteria of fairness and justice

Pareto optimality, first formulated by Pareto in 1909 (Pareto, 1909), is still the fundamental goal of neoclassical welfare economics. It defined allocation of resources Pareto efficient if there is no other feasible allocation which will make everyone better off (or, as more usually stated, make everyone at least as well off and at least one member better off). In other words, it says that a state in which no one can be made better off without making someone else worse off is Pareto optimal.

As already presented in Chapter 9, since the original development of DEA, Pareto-Koopmans efficiency is a fundamental property used in DEA. In the DEA literature, there are two efficiency notions: weak efficiency and Pareto-Koopmans efficiency. The Farrell-Debreu measure is calculated based on the weakly efficient frontier and hence possibly existing nonzero slacks are ignored in its efficiency measurement. By contrast, the original model by Charnes et al. (1978) is developed with the intension of incorporating the notion of full efficiency or Pareto-Koopmans efficiency.

Standard or black-box DEA is a set of mathematical programming techniques for measuring the efficiency performance of DMUs that convert exogenous inputs into final outputs. In standard DEA, the internal production processes of DMUs are ignored and the exogenous inputs consumed and final outputs produced by the DMUs are the only consideration for efficiency evaluation.

Charnes, Cooper, and their associates have incorporated this efficiency notion with the use of the non-Archimedean infinitesimal. In a black-box setting, a DMU is Pareto-Koopmans efficient if and only if it is impossible to make an improvement in the utilization of any input or output without worsening some of the other inputs and/or outputs. Hence, Charnes, Cooper, and their associates relate the CCR model to the notion of Pareto-Koopmans efficiency. See Charnes and Cooper (1984, 1985), Charnes, Cooper, Golany and Seiford (1985), and Cooper, Seiford, Tone and Zhu (2007) for detailed discussion of the difference between the two notions.

In his seminal article of "The Measurement of Productive Efficiency," Farrell started the introduction identifying the importance of efficiency analysis for policy making (Farrell, 1957):

The problem of measuring the productive efficiency of an industry is important to both the economic theorist and the economic policy maker. If the theoretical arguments as to the relative efficiency of different economic systems are to be subjected to empirical testing, it is essential to be able to make some actual measurements of efficiency. Equally, if economic planning is to concern itself with particular industries, it is important to know how far a given industry can be expected to increase its output by simply increasing its efficiency, without absorbing further resources.

Using the "activity analysis concept," Farrell connected deficiencies in commonly used index *number approaches to productivity (and like) measurement.* (Cooper et al., 2004a, b).

Jointly with Leonid Vitaliyevich Kantorovich, in 1975 Tjalling Koopmans shared a Nobel Prize "for their contributions to the theory of optimum allocation of resources."[13] Building on the work on Tjalling Koopmans' "Activity analysis Concept" (Koopmans, 1951), Charnes, Cooper, and Rhodes extended Farrell's work and succeeded in establishing the DEA as a basis for efficiency analysis.

In their original study, Charnes, Cooper, and Rhodes (1978) extended the $\frac{Outputs}{Inputs}$ to multiple inputs and multiple outputs, described DEA as a "mathematical programming model applied to observational data [that] provides a new way of obtaining empirical estimates of relations—such as the production functions and/or efficient production possibility surfaces—that are cornerstones of modern economics."

DEA's empirical orientation, it's easily used methodology for modeling operational processes for performance evaluations, and the absence of a need for the numerous a priori assumptions that accompany other approaches provided the opportunity for its wide application in a number of studies involving efficient frontier estimation in the governmental and nonprofit sector, in the regulated sector, and in the private sector.

To better connect with Buchanan framework, we shall recall the bellow definition of relative efficiency as applied in DEA, that, as already noted, has the advantage of avoiding the need for assigning a priori measures of relative importance to any input or output:

NO GAIN- NO LOSS: Definition 16.1 Efficiency — Extended Pareto-Koopmans Definition: Full (100%) efficiency is attained by any DMU if and only if none of its inputs or outputs can be improved without worsening some of its other inputs or outputs.

[13] https://www.nobelprize.org/prizes/economic-sciences/1975/koopmans/facts/.

With an understanding that in most management or social science applications, the theoretically possible levels of efficiency will not be known, the above definition is replaced by emphasizing its uses only with *the empirically available information*:

WIN- NO LOSS: Definition 16.2 Relative Efficiency: A DMU is to be rated as fully (100%) efficient on the basis of available evidence *if and only if the performances of other DMUs does not show that some of its inputs or outputs can be improved without worsening some of its other inputs or outputs.*

Here again, the definition does not need prices or other assumptions of weights which are supposed to reflect the relative importance of the different inputs or outputs and does not require an explicit specification of the supposedly existing formal relations between inputs and outputs.

Various models with associated measures have been devised to eliminate this weakness in the Farrell model. An example is the "additive model" which was introduced in Charnes et al. (1985). This model can be represented by setting $\phi^* = 1$ in (9.6a). The conditions for efficiency set forth in *Definition 2* presented in Chapter 9, is reformulated as follows:

WIN-WIN: Definition 16.3: *Full or Strong (Pareto-Koopmans) efficiency is attained for DMUo by an additive model if and only if all slacks are zero in an optimum solution. That is, if and only if* $s_i^{-*} = s_r^{+*} = 0 \quad \forall i, r$.

Discarding condition (1) in Definition 2 eliminates the distinction between inefficiencies that can:

(1) be eliminated without altering the input (mix) proportions and

(2) extends the measure of inefficiencies to include nonzero output slacks.

It is noteworthy, however, that contact between the "CCR" (=Charnes, Cooper, and Rhodes) model that is represented in (9.6a) is maintained via the following:

Theorem 1: The additive model will identify a DMU as efficient if and only if the CCR model represented in (9.6a) also identifies it as efficient. That is, full (or strong) efficiency is to be achieved in both cases.

This implies that efficiency in the additive model is sufficient but not necessary for efficiency in the Farrell model. This theorem is proved in Ahn, Charnes and Cooper (1985) so we need not prove it here. Instead, we note that the additive model does not use a ratio form, like in the Farrell model.

While Buchanan recognizes that "the Pareto criterion is of little value when employed solely to classify 'results' defined with respect to the orthodox economic variables, inputs and outputs of goods and services possessed by different persons in the social group" he also suggests, "the criterion must be extended to classify social rules which constrain the private individual behavior that produces such result."

The fair society approach to development recommends appraising policy interventions according to how effective they are likely to prove in facilitating voluntary cooperation among individuals in their pursuit of their own social welfare and happiness. In our studies, where the DEA is applied to the policy and performance indicators of countries, the interpretation of *Pareto-Koopmans* efficiency rises up to *effectiveness* considerations and searches to identify WIN-WIN; WIN- NO LOSE *policy packages* that yield gain for all/some and loss for none that could command unanimous consent. Within our framework, the unanimous consent is not expected for each separate feature of a policy change, but only for the entire multiple objective package. Insofar as no policy changes can be found in the real world that would harm literally no one dimension, then, in a corresponding sense, the existing combination of policies is Pareto-optimal.

A distinguishing feature of DEA is that it can handle multiple noncommensurate input/output variables interacting in complex and unknown ways. The DEA is a "black-box" technique used when there is insufficient knowledge of causal relationships; or there is a lack of consensus on the alternatives. It is capable of addressing managerial, social, and economic issues and provides useful results for conducting *ex-post* evaluations of performances of decision/policy-making units (PMUs) that enable decision makers to develop *ex-ante* measures directed to improving future performances. DEA is also a benchmarking tool for identifying the best production/transformation practices.

It thus provides a new analytical framework for evaluating *ex-post* performances and accomplishments of countries in their socioeconomic and environmental development and provides a basis for *ex-ante* planning of future activities. See Emrouznejad, Parker, and Tavares (2008) who list more than 4200 papers and book chapters by 2500 authors that treat a wide variety of uses and developments of DEA (including country and regional development evaluations) in more than 42 different countries. See also Cooper, Seiford, and Tone (2007) and Cooper, Seiford, Thanassoulis and Zanakis (2004b); Seiford and Thrall (1990).

In view of the above, we suggest it to become a powerful analytical tool for evaluating the effectiveness of various policy packages on the progress of countries and other DMUs toward the achievement of SDGs and ensuring "shared prosperity."

One of questions researchers ask, is if the DEA-based models belong to the family of global optimization models? In this regard, it is noteworthy to recall that *global optimization* is a branch of applied mathematics and numerical analysis that attempts to find the global minima or maxima of a function or a set of functions on a given set of restrictions. It is usually described as a minimization problem because the maximization of the real-valued function $G(x)$ is equivalent to the minimization of the function $F(x) := -G(x)$.

It is noteworthy that the main constraints inherent in DEA models tie all nations together, when applied to cross country analysis. The SDG welfare function of a nation determined by the optimization program does not just reflect the goal preferences of that nation alone. The solution weights of the goals emerge as the result of an overarching optimization that includes information about every other nation, and provides solutions that can guide *team consensus* about the priorities of the goals.

10.8 Toward a new social contract for recovery and resilience

Along with remarkable progress in the fight against poverty and technological progress, the globalization failed to deliver the much aspired positive equitable transformation. Rather, they are attributed the persistence of widespread informalities and multiple deficits in terms of fiscal sustainability and decent work, a gap of 50% in food, 45% in energy, and 30% in water consumption to be filled by 2030, etc.

The current Covid crisis serves as a manifest proof to that real world governments not only failed to adopt the optimal multidimensional solutions, but often intervened in ways that promoted economic inefficiency, thereby clashing sharply with the economists' prescriptions for equitable development. There is a universal conviction that for reversing the damage, it is the time to revisit the current development paradigms by seeking the ways for assuring sustainable advancement of world economy through managing the globalization maximizing it's benefits and reducing it's negative impact. The time has come to inverse the trajectory, by putting sustainable development at the core of decision making and to revisit the role of macroeconomic governance: *create policies conducive to a more equal, prosperous, peaceful and just world.*

To cope with widespread market failures and asymmetries, a new paradigm is needed, one that will ensure a balance between markets and the constructive government policies directed to facilitating structural transformations, diversification, and industrial upgrading. A new development model is needed, one that will enhance efficiency, effectiveness, transparency, accountability by means of creating a new "Geneva Consensus" as a new basis to manage globalization accounting for the resultant multidimensional adjustment costs, as inspired by former WTO Director General Pascal Lamy.

The Agenda 21 along with the outcome document of the UNCSD Rio+20 "The future we want" can be named among the UN achievements in spelling out the commitment of the international development community toward the common cause of sustainable future. The need for a new global partnership and dialog that can facilitate the process of negotiating compromises between stakeholders has emerged.

Achieving sustainable development calls for *synergies and coherence* of these three pillars, based on principles of democracy, solidarity, and gender justice. Consequently, the only legitimate means to ensure transparency and ownership over the development process are interconnected mechanisms of policy dialog and consensus building. We therefore argue that the social capital of a united community is the catalyst for successful and sustainable change. Attainment of sustainable development is a social movement toward eliciting consensus that will lead to concluding and implementing a social contract much inspired by Jean-Jacques Rousseau.

There is a firm consensus that international trade, financial and monetary systems should be reformed to propose innovative *integrated, coordinated, and coherent* policy packages for managing globalization that, based on the principles of generational accounting, would assure the sustainable advancement of the world economy and simultaneous pursuit of SDGs in developing and developed countries alike.

The time to move from vision to action has arrived! Building on the historical achievements of MDGs and Rio +20, the main target of the global community is to elaborate and implement "An Action Agenda for Sustainable Development." In the wake of this transformative shift, the core task is to address the hard issue of creating an integrated policy framework within the UN system and at the global, regional, and national levels, with due respect to countries' ownership and grounded in respect for universal human rights, peace, and security. Mechanisms for participatory decision-making processes that will manage effectiveness-efficiency-equity trade-offs in globalization need to be developed.

With a widespread recognition that a narrow, GDP-based traditional measure of societal well-being, has become obsolete a new *dynamic and diagnostic metrics* of societal well-being is needed, that will go "beyond GDP" to a "triple bottom-line" of economic, social, and environmental performance and will recognize peace and security as a core element of well-being and good governance.

We therefore suggest the DEA to become a strategic means of participatory decision-making process that will treat trade-offs and will enable all development stakeholders to make informed choices and build a consensus on managing the process of globalization and plan future development policies of countries toward the achievement of SDGs. We submit that, only within a new global partnership will the global world be able to strike a balance and create the foundations for innovative policy packages ensuring convergence of the well-being of people, planet, and societal prosperity. We build on accepted perceptions that:

- Understandings between developing and developed countries play a key role, both within country groups as well as in global society.
- Through regional and global dialog, a more inclusive form of globalization can emerge.

Consensus is the foundation of stability and sustainability, forging an international bipolar, triangular, and global consensus is a potent mechanism for shaping national and international policies that are socially just, economically productive and environmentally friendly.

Our preliminary results in evaluating the social and economic performance of countries with the application of the DEA reported in Part 1, and our earlier work, empirically confirm that the world is still far from realizing the full potential of globalization. However, *if* wisely *managed*, globalization *can* deliver unprecedented material progress, generate more productive and better jobs, and contribute significantly to reducing poverty. Thus, we find that: a better world is *possible*; the benefits of globalization *can* be extended to *more* people and countries; the potential for good is *immense* and *can* be achieved in the context of *open* economies. It is remarkable that DEA was applied in so many different domains. With tremendous increase in applications ranging from banks to countries, the potential of DEA needs to be unlashed for the benefit of humanity.

Not only DEA adds a new dimension to existing theory of production/transformation analysis but it is capable of explaining a larger number of phenomena pertinent in cybernetic feedback processes and decision makers or regulators that is at the heart of systems science. The above properties are fundamental determinant of the traditional philosophy of a general science.

On the road to 2030, the international trade, financial and monetary systems should be reformed to propose innovative policy packages for managing globalization that, *based on the principles of generational accounting,* would assure the sustainable advancement of the world economy and simultaneous pursuit of SDGs in developing and developed countries alike:

– to better reflect the voting power of the changing global economy and to address persistent risks;
– to account for social and environmental *opportunity costs* and reconcile interests of capital and labor within planetary boundaries;
– to evaluate current and past achievements of policy and predict improvement opportunities;
– to identify the *efficient-effective-equitable* allocation mode for Earth's finite natural resources.

The Rio +20 outcome document "The Future We Want" resolved to establish an inclusive and transparent intergovernmental process on SDGs that is open to all stakeholders with a view to developing global sustainable development agenda. The implementation of Agenda 2030 process leading to the SDGs needs to be integrated, coordinated, and coherent.

Implementation of an integrated approach to poverty reduction and sustainable development, in this age of globalization, requires policy coherence and the political will to create a strong partnership between national and international stakeholders, within their fields of competences and constitutional objectives of their respective institutions (Figs. 10.1 and 10.2). In Fig. 10.1, we have made an attempt to present a SDGs Implementation Framework, based on the principles of structural, inclusive and systematic approach, that we believe will ensure 3Cs- Coherence, Coordination and Cooperation among all development stakeholders. We explicitly recognize the country ownership over the implementation of Agenda 2030, with technical and policy support from international development community in the implementation of Country's Action Agenda for Sustianble Development, based on their institutional mandates and constitutional objectives, for helping countries' achieve SDGs.

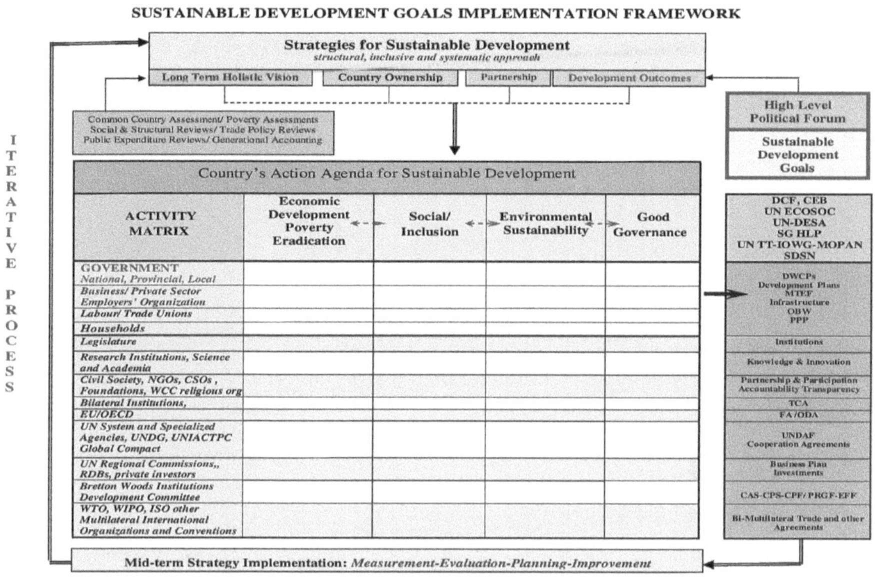

Figure 10.1 Coherence, coordination, and cooperation for achieving SDGs.
CAS, World Bank's Country Assistance Strategy for Low-Income Country (LIC);
CEB, The UN Chief Executive Board;
CPF, World Bank's Country Partnership Framework;
CPS, World Bank's Country Partnership Strategy for Middle Income Country (MIC);
DCF, Development Cooperation Forum;
DWCP, Decent Work Country Programme;
EFF, International Monetary Fund's Enhanced Fund Facility for MIC;
FA, Framework Agreements;
HIPC, Heavily Indebted Poor Countries Initiative;
IOWG, Intergovernmental Open Working Group;
MDRI, The Multilateral Debt Relief Initiative;
MOPAN, Multilateral Organization Performance Network;
MTEF, Medium-Term Expenditure Framework, prepared by government;
OBW, Outcome based Work-plan;
ODA, Official Development Assistance;
PPP, Public Private Partnerships;
PRGF, International Monetary Fund's Poverty Reduction and Growth Facility for LIC;
PRSP, Poverty Reduction Strategy Paper;
RDB, Regional Development Bank(s);
SDSN, Sustainable Development Solutions Network;
SG HLP, The Secretary General's High Level Panel on the post-2015 Agenda;
TCA, Technical Cooperation Agreement;
UN-DESA, UN Department of Economic and Social Affairs;
UN ECOSOC, United Nations Economic and Social Council;
UNDAF, United Nations system's Development Assistance Framework;
UNDG, The UN Development Group;
UNIACTPC, The United Nations Inter-Agency Cluster on Trade and Productive Capacity;
UN TT, UN System Task Team on the post-2015 UN Development Agenda;
WCC, World Council of Churches.

UNDP:	1997, Human Development Report noted that "If globalization is to afford greater opportunities for developing countries, it needs more careful management, and must take into account global equity concerns."[14]
UNCTAD:	1997, The 44th session of UNCTAD's Trade and Development Board expressed a view that Globalization needs international regulations as "the globalization process is making the national economies of both developing and industrial countries more difficult to manage and international regulations are needed to enable developing countries to maximize the benefits and minimize the costs of integration into the world economy".[15]
World Bank:	2003, most people agree that globalization has both positive and negative effects; and that the world lacks good institutions to ameliorate the negative ones. The ***"Equator Principles,"*** created by the private-sector arm of the World Bank in 2003 and embraced by 40 big private banks, were a rare creative effort to grapple with this deficit.[16]
IMF:	2005 The IMF, along with the World Bank, has contributed significantly to the Monterrey conference by helping to focus the conference on global priorities, such as the Millennium Development Goals. The IMF Managing Director, Rodrigo de Rato, during his Speech on October 20, 2005 recognized that "Reversing the process of globalization would not solve the problem of poverty—that was amply demonstrated by the events of the 20th century. The world needs instead a new approach to globalization that exploits its enormous potential for improving human welfare. In order to carry the process forward, and build support for a better globalization, a common political understanding of how to maximize the benefits, while minimizing the risks, must be developed. He provided a conceptual framework for IMF/World Bank Involvement in Globalization, recognizing that "the IMF is part of a wider network of international institutions that each has an important role to play in making globalization work better, stated: I would like the main focus of the Fund's work in the next few years to be on helping our members maximize the benefits, and minimize the costs and risks, of globalization."[17]
OECD:	2005 The OECD Ministerial Council Meeting reaffirmed the strong commitment to the implementation of the Millennium Declaration and the Monterey Consensus and for achieving the MDGs: "We are engaged also in internationally coordinated endeavours to ensure that development progress is properly measures and assessed and that results are used for improving development performance and aid effectiveness." [OECD, C/MIN(2005)]
UN:	2006, The UN Secretary General, in his speech at the 4th European Union/Latin America and Caribbean Heads of State Summit in Vienna stated "First, and foremost, there is an urgent need to prioritize employment in decision making. Traditional policy discussions treat job creation as an inevitable outcome of economic growth. As a result, economic policy formulation has focused more on keeping inflation in check and increasing output than creating employment. Yet there is mounting evidence that growth alone, while crucial for employment, does not always lead to enough jobs... We must re-evaluate our approach, and place job creation right next to economic growth in national and international economic and social policies. For instance, when discussing macroeconomic policies there should be an institutionalized reflex which constantly asks "what can this do for jobs?"[18]
WTO:	2006 Director-General Pascal Lamy, called for a creation of a new "Geneva consensus": "a new basis for the opening up of trade that takes into account the resultant cost of adjustment. "We need to "reform globalization" with a clear view to enhancing the development of social, economic and ecological aspects of humanity. This is also in line with the millennium development goals that can be achieved through a "reform of globalization from within and for development". [Lamy 2006] Trade opening is necessary, but it is not sufficient in itself. Humanizing globalization means that we must take care of the victims of globalization. To achieve this, global solutions must be sought for addressing the negative impact of globalization at all levels — individual, community and universal levels. "The powers of governments to intervene to minimise costs and maximise benefits for host communities and society are being eroded by bilateral investment treaties and the proposed investment agreement in the WTO"[19]
WHO:	2006 WHO in turn underscores the importance of policy coherence "We are working to ensure that trade and globalization work to improve the health of the poor and disadvantaged populations. The overall goal is to achieve gre ater coherence between international trade and health policy so that international and multilateral trade rules maximize health benefits and minimize health risks especially for poor and vulnerable populations."[20]
IFC:	2006 To promote sustainable private sector investment and increase the development impact of their projects, in a globalizing world, the Board of Directors of the International Finance Corporation adopted new ***environmental and social standards for the organization***. The new standards build upon the environmental and social requirements that IFC currently applies to private sector projects it finances in the developing world.[21]

Figure 10.2 Optimal management of globalization: Maximizing benefits and minimizing costs.[14–21]

[14] TWN, Third World Network, http://www.twnside.org.sg/title/mor-cn.htm.

[15] TWN, Third World Network, http://www.twnside.org.sg/title/regu-cn.htm.

[16] http://psdblog.worldbank.org/psdblog/2006/02/responsibly_man.html.

[17] Globalization: A Framework for IMF Involvement. http://www.imf.org/external/np/speeches/2005/102005a.htm.

[18] http://www.un.org/apps/sg/sgstats.asp?nid=2030.

[19] Trade liberalisation statistics, http://www.gatt.org/trastat_e.html.

In a search for a fair and sustainable globalization, all major international organization and conferences, reached a consensus on a need of maximizing favourable impact of globalization and minimizing its undesirable consequences and mitigating the social costs Fig. 10.2. The proposed framework can be used to deliver the analytical underpinnings needed by the global decision makers in order to take informed policy decision on ensuring policy coherence with regard to the optimal levels sustainability impact of globalization. It will also facilitate the harmonic convergence of the complex relationship between economic development, social progress, and environmental protection within a broader inclusive framework at various levels of decision making.

"Geneva Consensus" will provide the platform for fostering synergies on key policy issues and pursuing a collaborative framework aimed at mobilizing the World Community's potential to develop improved, proactive, integrated responses to the global economic turmoil.

These include governments, legislators, employers, workers, and the financial system and also academics and civil society representatives:

- to set common goals,
- to create enabling institutional capacity, social, financial, and physical infrastructure,
- to harness the ingenuity and dynamism of business for sustainable development,
- to form global partnership for development data,
- to develop evidence-based solutions,
- to mobilize action and resources.

Developing countries should increase their productivity and enhance competitiveness by creating new socially responsible, resource-energy-efficient and environmentally friendly productive capacities. The Report of the UN Secretary General's High-Level Panel on Post 2015 Development Agenda: "A New Global Partnership" contends that a new spirit of solidarity, cooperation, and mutual accountability must underpin the post-2015 agenda:

- to drive five fundamental transformative shifts,
- to build multi-stakeholder partnerships,
- to forge a political consensus.

Globalization does, indeed, provide a central vehicle for a broad-based and sustainable prosperity and is vital for helping countries to transfer modern green technologies and innovative solutions, and to restructure their current production systems away from Business as Usual practices toward sustainable trajectory. It is also important that developed countries honor their aid commitments and harmonize their procedures with national plans, and through effective and efficient development cooperation and development finance channeled through the government budgets, assist developing countries' efforts in moving to a position where they can mobilize their own resources for sustainable development.

The ILO Constitution stresses that "poverty anywhere constitutes a danger to prosperity everywhere." With a significant change in the mindset and traditional way of looking at the balance-sheet, today's about 700 million poor can be moved from the world's debit accounts to its credit accounts. They can then be viewed as a potent source for increased aggregate demand and long-term sustainable global growth, contributing to a win-win solution toward sustainable world economy.

As noted earlier, from an analytical perspective, attainment of sustainable globalization calls for *Pareto-Koopmans-efficient solutions*. Delivering on UN declarations on integrated analysis of SDGs here we proposed an *interactive-iterative policy evaluation system for monitoring progress made toward achievement of SDGs*. Unlike the deliberations around Washington Consensus, the Geneva Consensus advocates multidisciplinary, multiple-level, and cross-sectoral integrated systems approach to global governance. We presume global decision making that engage all development stakeholders in all phases of the process of addressing existing interlinkages among the SDGs, to be able to bring the information related to their own concerns to the discussion table and thus to ensure informed and evidence-based policy formulation. We call to enlist the support of interdisciplinary group of specialists: not only economists but also lawyers, sociologists, anthropologists, and others. Public-private partnerships can also help scale up and speed up access to resources, specialized skills, innovations, and the know-how necessary to build, operate, and maintain environmentally sustainable and people-centered globalization.

The availability of statistical information has considerably improved during the implementation of the MDGs and recent years. The advent of information and communications technologies (ICTs) along with the fourth industrial revolution provides an unprecedented opportunity to strengthen data collection and statistical analysis. Enhanced data availability will create the information basis for development of global models, thus providing an analytical underpinning for global governance. Along with "A World that Counts: Mobilizing the Data Revolution for

Sustainable Development" World Summit on the Information Society Forum 2017 underscored the important role of ICT for realizing the SDGs and will create legal, political, and methodological grounds for a change and countless empirical simulations so much desired by Coase, Kydland, and others.

The classical change management teaches as to take change as an opportunity and I believe the current sanitary crisis that has shacked the foundation of all our believes, is an unprecedented opportunity for a major change in the way we think about the development agenda.

For the first time since 1945, in the entire history of the existence of the United Nations, it highlights strongly the value and the power of the very first paragraph of Preamble to the United Nations Charter:

We the Peoples of the United Nations Determined:

- *to save succeeding generations from the scourge of war, which twice in our lifetime has brought untold sorrow to mankind, and*
- *to regain faith in fundamental human rights, in the dignity and worth of the human person, in the equal rights of men and women and of nations large and small, and*
- *to establish conditions under which justice and respect for the obligations arising from treaties and other sources of international law can be maintained, and*

And for these Ends

- *to practice tolerance and live together in peace with one another as good neighbors, and*
- *to unite our strength to maintain international peace and security, and*
- *to ensure, by the acceptance of principles and the institution of methods, that armed force shall not be used, save in the common interest, and*
- *to employ international machinery for the promotion of the economic and social advancement of all peoples.*

The Covid crisis imposes on us a need to reframe the conventionally defined development agenda to a 21st century global social contract in the context of globalization, one that will maximize the benefits of global economic interdependence while minimizing the risks and costs already stipulated in a large number of occasions. We need to forge a new global deal that will serve the interests of all societal groups, not only for the world's poor but for the world's indispensable middle class but also those vulnerable groups who have found themselves in poverty traps of in the rich world's mature democratic economies as identified in the Part 1 of our book. A new sustainable deveopment agenda that will promote social progress and better standards of life in larger freedom for all is badly in need.

It is widely accepted, that by the virtue of a definition sustainable development as a global imperative uniting the common interest of all nations and people, developed and developing alike, the Agenda 2030 has already laid the formal ground for the formation of such a global social contract. With the tiny virus, the long perceived distinctions between the world's rich people and powerful nation states and the poor and unprotected has evaporated. The pandemic came to confirm the highest level of interdependence on prosperity and security in the global village, highlighted the need of global risk management, early warning and insurance facilities, the new imperative for innovative approaches for ensuring universal social protection, and the development of so much inspired universal social security system.

As already discussed, the application of DEA models at the global level, within the context of the public choice revolution provides the answer for a convergence of global political priorities with the longstanding development agenda. Government is a set of participatory decision-making processes by which people and major national and international development stakeholders relate to one another. Governmental decisions and policies are based within a particular political system or constitutional order. Agenda 2030 is a perfect example of such political system encompassing the set of SDG objective that raise to the level of the constitutional order. The political process of development and implementation of "SDGs Implementation Framework," can be visualized as a complex set of interrelationships among all stakeholders at various levels: local, regional, national, and global, inscribing each of them citizens, politicians, and bureaucrats specific set of responsibilities within their fields of competences and constitutional objectives of their respective institutions.

10.9 Why "Geneva Consensus"?

Consensus has long been prized and typically founded upon moral and scientific arguments, where the interests of all stakeholders involved in establishing the consensus are backgrounded and respected.

The name "Geneva Consensus," assigned to the foundation, that I have co-founded in January 2013, to advocate the achievement of SDGs in the context of globalization, is found very appealing by many. However, hardly anyone sees the profound scientific foundation behind the purely declarative nature of the name of our NGO.

The Geneva Consensus, as we apply it here, implies the welfare-political-economist paradigm and marries the social welfare economics considerations with political economy. Joining a fair globalization militants' group, in their quest for a paradigm shift, we created thc "Geneva Consensus Foundation" a new institutional basis, to contribute to managing globalization, aiming to provide an analytical underpinning that is adept to enhance efficiency, effectiveness, transparency, and accountability of the global world to account for the resultant multidimensional adjustment costs.

It is noteworthy, that already in my 2012 article (Tarverdyan, 2012), published in UN Special N^0 721 October edition (Tarverdyan, 2012), entitled "Geneva Consensus For Sustainable Development: Toward a New Social Contract," I have submitted that from an analytical perspective, attainment of sustainable globalization calls for Pareto-Koopmans-efficient solutions. Created in 1949 by the UN Geneva Office Staff Coordinating Council and the WHO Staff Committee, UN Special is the monthly magazine that aims to keep the UN community abreast of what is going on at the UN.

I have noted that delivering on UN declarations on integrated analysis of SDGs we propose *an interactive-iterative policy evaluation system for monitoring progress made toward achievement of SDGs*. And so already then, I presented the DEA to become *a new interactive-iterative performance assessment and consensus building framework* for evaluating the multidimensional development of *decision-making units* (enterprises, schools, banks, cities, regions, countries, etc.) toward the achievement of, then yet nonexistent SDGs.

Our empirical analysis demonstrates that most of the countries record multiply deficits: "Sustainable Development Deficit", "Decent Work Deficit" and "Fairness Deficit of Globalization" we also derive numerical estimates of such deficits. To reverse the damage it is the time to revisit the current development paradigms and seek ways to ensure sustainable advancement of the world economy. The analytical approach presented in this book, entails the following advantages:

- Evaluation of the performance and competitiveness of units such as nations, regions, cities, projects, enterprises banks, etc., transcending well "beyond GDP" standard measures to more comprehensive economic, social and environmental attributes;
- Providing a basis for integrated packages of policy advice for comprehensive industrial, regional, or global agendas;
- Establishing surveillance instrument for tracking development progress, as well as a consensus-building tool contributing toward increasing benefits and reducing social-environmental costs of globalization.

There is a clear recognition that the international institutional frameworks for sustainable development are poorly developed. Given sustainable development's multidisciplinary and crosscutting nature, its attainment is conditioned by the ability to form global partnerships tapping into the knowledge and raising commitment of the international development community particularly the WTO-ITC, UNCTAD, UNEP, World Economic Forum, ILO, UNECE, and other UN Agencies and International Organizations—all based in Geneva. In partnership with the World Bank, IMF, OECD, and other bi/multilateral international organizations, Geneva is best-placed for providing a new forum for dialog on key policy issues with the participation of global decision makers and for monitoring the performance under agreed international agendas.

The term "consensus" is derived from the Latin cōnsentiō ("feel together; agree"), a fusion of cōn- ("together") and sentiō ("sense; perceive; feel") and the neutrality is the prime precondition to either one or the other.

Switzerland has historically been a global center for accommodating fair and neutral outcomes on the most sophisticated political decisions in the spirits of solidarity. Neutrality is a foreign policy norm for Swiss authorities, firmly stipulated in the Federal Agreement of 1815 and in the constitutions of 1848, 1874, and 1999. Swiss neutrality has served as a model for others.

At national level, neutrality is cited in the Federal Constitution as an instrument for safeguarding independence. The essential rights and duties of a neutral state are laid down in the Hague Conventions on Rights and Duties of Neutral States signed in 1907. Switzerland is one of the 33 members of the Hague Conventions and in 1920 joined the League of Nations, and Geneva became the seat of the League. In 1939, Switzerland formally declared its neutrality in accordance with the Hague Conventions. Swiss perceive a power to be "neutral when it does not take sides in a war."

What is commonly considered as risks is the impact of the environment within the systems thinking and it reciprocal influence on the system. Within that context, the neutrality of Switzerland ensures that the global decision making is liberated from any such risk.

Switzerland is the first country whose membership of the UN was decided directly by its people by a referendum in 2002. In its accession statement, Switzerland affirms: "Switzerland is a neutral state whose status is based on international law. Even as a member of the UN, Switzerland remains neutral." Switzerland and the UN pursue the same objective: to promote and guarantee peace and security in the world. It is therefore in Swiss interests to support the UN through its policy of active neutrality and practical solidarity.

I should like to conclude with the words of Jean Jacques Rousseau in 1754 in *Discourse on Inequality* that he dedicated to the state of Geneva, he praised Geneva as a good, if not perfect, republic that ensures stability of its laws and institutions, the community spirit of its inhabitants, and its good relations with neighboring states, neither threatening them nor threatened by them, and the well-behaved women of Geneva.

It is, perhaps, given to the city of Geneva alone, to produce the edifying example of so perfect a union between its clergy and men of letters. It is in great measure on their wisdom, their known moderation, and their zeal for the prosperity of the State that I build my hopes of its perpetual tranquility.

References

Abad, C., Thore, S., Laffarga, J., 2004. Fundamental analysis of stocks by two-stage DEA. Manag. Decis. Econ. 25, 231–241.

Adams, B., Judd, K., 2018, April 26. The Ups and Downs of Tiers: Measuring SDG Progress. Global Policy Watch (Also available in Spanish).

Adler, N., Golany, B., 2007. Data reduction through principal component analysis (DEA-PCA). In: Cook, W., Zhu, J. (Eds.), Modeling Data Irregularities and Structural Complexities in Data Envelopment Analysis. Springer, New York, NY, p. 52.

Adler, N., Yazhemsky, E., 2007. Improving Discrimination in Data Envelopment Analysis: PCA-DEA versus Variable Reduction. Which Method at What Cost? Working Paper. Hebrew University of Jerusalem, Jerusalem, p. 54.

Adler, N., Yazhemsky, E., Tarverdyan, R., 2010. A framework to measure the relative socio-economic performance of developing countries. Socio-Econ. Plan. Sci. 44 (2), 73–88.

Afonso, A., Aubyn, M.S., 2005. Non-parametric approaches to education and health efficiency in OECD countries. J. Appl. Econ. 8, 227–246, 55.

Afsa, C., Blanchet, D., Marcus, V., Mira d'Ercole, M., Pionnier, P.A., Ranuzzi, G., Rioux, L., Schreyer, P., 2008. Survey of Existing Approaches to Measuring Socio-economic Progress. Working Paper. Commission on the Measurement of Economic Performance and Social Progress, Paris.

Akther, S., Fukuyama, H., Weber, W.L., 2013. Estimating two-stage network slacks-based inefficiency: an application to Bangladesh banking. Omega 41 (1), 88–96.

Ali, I.A., Seiford, L.M., 1990. Translation invariance in data envelopment analysis. Oper. Res. Lett. 9 (6), 403–405.

Arrhenius, S., 1896. On the influence of carbonic acid in the air upon the temperature of the ground. Philos. Mag. J. Sci. 41, 237–276.

Arrow, K.J., 1958. Tinbergen on economic policy. J. Am. Stat. Assoc. 53, 89–97.

Arthur, W.B., 1989. Competing technologies, increasing returns, and lock-in by historical events. Econ. J. 99, 116–131.

Ashby, R., 1962. Principles of the self-organizing system. In: Principles of Self-Organization: The Illinois Symposium on Theory and Technology of Self-Organizing Systems. Pergamon, London.

Ashby, R., 1964. In: Ross Ashby, W. (Ed.), An Introduction to Cybernetics. Methuen London.

Ashby, R., Stein, P.R., 1954. Design for a brain. Phys. Today 7 (4), 24.

Avdagic, S., Rhodes, M., Visser, J., 2011. Social Pacts in Europe: Emergence, Evolution, and Institutionalization. Oxford University Press, Oxford.

Balassa, B., 1989a. Comparative Advantage, Trade Policy and Economic Development. Harvester Wheatsheaf, New York, NY.

Balassa, B., 1989b. New Directions in the World Economy. Macmillan, New York, NY.

Bana e Costa, C., Thore, S., 2002. The many dimensions of an R&D project: multi-criteria analysis. In: Thore, S. (Ed.), Technology Commercialization: DEA and Related Analytical Methods for Evaluating the Use and Implementation of Technical Innovation. Kluwer, Boston, MA, pp. 23–44.

Bana e Costa, C., de Corte, J.M., Vansnick, J.-C., 2012. MACBETH. Int. J. Inf. Tech. Dec. Mak. 11, 359–387.

Bandura, R., Saisana, M., 2012. An Inventory of 300 Composite Indicators. European Commission, JRC-IPSC, Italy (No EUR 25560 EN).

Banker, R.D., Thrall, R.M., 1992. Estimation of returns to scale using data envelopment analysis. Eur. J. Oper. Res. 62 (1), 74–84.

Banker, R.D., Charnes, A., Cooper, W.W., 1984. Some models for estimating technical and scale inefficiencies in data envelopment analysis. Manag. Sci. 30, 1078–1092.
Barbier, E.B., Burgess, J.C., 2019. Sustainable development goal indicators: analyzing trade-offs and complementarities. World Dev. 122, 295–305.
Barr, N., Economics of the Welfare State, 2012. University Press, Oxford.
Bass, F.M., 1969. A new product growth model for consumer durables. Manag. Sci. 215–27.
Bateson, G., 1972. Steps to an Ecology of Mind: Collected Essays in Anthropology, Psychiatry, Evolution, and Epistemology. Chandler Publishing Company, San Francisco, CA.
Ben-David, D., Nordstrom, H., Winters, A., 2001. Trade, Income, Disparities and Poverty, Special Studies 5. World Trade Organization, Geneva.
Bennich, T., Weitz, N., Carlsen, H., 2020. Deciphering the scientific literature on SDG interactions: a review and reading guide. Sci. Total Environ. 728, 138405.
Billström, F., Thore, S., 1964. Simulation experiments with dynamic price strategies in monopoly theory. In: Wold, H. (Ed.), Econometric Model Building. North-Holland, Amsterdam, pp. 283–296.
Bird, K., 2008. How Development Cooperation Can Support Pro-poor Policy Processes. ODI, London forthcoming.
Birdsall, N., 2008. The Development Agenda as a Global Social Contract, or, We Are All in This Development Boat Together. Lecture given at the Dutch Scientific Council (WRR), The Hague. December 08, 2008.
Boulding, K.E., 1941. Economic Analysis. Harper & Brothers, New York, NY.
Bouyssou, D., 1999. Using DEA as a tool for MCDM: some remarks. J. Oper. Res. Soc. 60, 974–978.
Buchanan, J.M., 1959. Positive economics, welfare economics, and political economy. J. Law Econ. 2, 124–138.
Buchanan, J.M., 1964. What should economists do? South. Econ. J. 30 (3), 213–222.
Buchanan, J.M., 1979. Public choice and public finance. In: What Should Economixsts Do? Liberty Press, Indianapolis, IN.
Buchanan, J.M., 2003. The constitutional way of thinking. Supreme Court Econ. Rev. 10, 143–155.
Bukenya, B., 2016. From social accountability to a new social contract? The role of NGOs in protecting and empowering PLHIV in Uganda. J. Dev. Stud. 52 (8), 1162–1176.
Bussolo, M., Davalos, M.E., Peragine, V., Sundaram, R. Toward a New Social Contract: Taking on Distributional Tensions in Europe and Central Asia. Europe and Central Asia Studies.
Bussolo, M., Lopez-Calva, L.-F., Sundaram, R., 2018. Leveling the Playing Field: Rethinking the Social Contract in ECA. The World Bank, Washington, DC.
Castelli, L., Pesenti, R., Ukovich, W., 2010. A classification of DEA models when the internal structure of the decision making units is considered. Ann. Oper. Res. 173, 207–235.
Chambers, R., 1983. Rural Development: Putting the Last First. Longman, Harlow.
Charnes, A., 1952. Optimality and degeneracy in linear programming. Econometrica 20, 160–170.
Charnes, A., 1953. Constrained games and linear programming. Proc. Natl. Acad. Sci. U.S.A. 639–641.
Charnes, A., Cooper, W.W., 1961. Management Models and Industrial Applications of Linear Programming, 2 vols. Wiley, New York, NY.
Charnes, A., Cooper, W.W., 1962. Programming with linear fractional functionals. Nav. Res. Logist. Q. 9, 181–196.
Charnes, A., Cooper, W.W., 1974. An extremal principle for accounting balance of a resource value-transfer economy. Rendiconti Accademia Nazionale dei Lincei 556–561.

Charnes, A., Cooper, W.W., 1977. Goal programming and multiple objective optimizations. Eur. J. Oper. Res. 1, 39–54.
Charnes, A., Cooper, W.W., 1984. The non-Archimedean CCR ratio for efficiency analysis: a rejoinder to Boyd and Färe. Eur. J. Oper. Res. 15, 333–334.
Charnes, A., Cooper, W.W., 1985. Preface to topics in data envelopment analysis. Ann. Oper. Res. 2, 59–94.
Charnes, A., Thore, S., 1966. Planning for liquidity in financial institutions: the chance-constrained method. J. Finan. 21, 649–674.
Charnes, A., Cooper, W.W., Mellon, B., 1952. Blending aviation gasolines: a study in programming interdependent activities. Econometrica 20, 135–159.
Charnes, A., Cooper, W.W., Ferguson, R.O., 1955. Optimal estimation of executive compensation by linear programming. Manag. Sci. 138–151.
Charnes, A., Cooper, W.W., Symonds, G.H., 1958. Cost Horizons and Certainty Equivalents: an approach to stochastic programming of heating oil. Manag. Sci. 4, 235–263.
Charnes, A., Cooper, W.W., Kozmetsky, G., 1973. Measuring, monitoring and modeling quality of life. Manag. Sci. 19, 1172–1188.
Charnes, A., Cooper, W.W., Rhodes, E., 1978. Measuring the efficiency of decision making units. Eur. J. Oper. Res. 2, 429–444.
Charnes, A., Cooper, W.W., Golany, B., Seiford, L., Stutz, J., 1983. Pareto-optimality, Efficiency Analysis and Empirical Production Functions, Research Report CCS, 479. Center for Cybernetic Studies, The University of Texas at Austin, Austin, TX.
Charnes, A., Cooper, W.W., Golany, B., Seiford, L., 1985. Foundations of data envelopment analysis for Pareto-Koopmans efficient empirical production functions. J. Econom. 30, 91–107.
Charnes, A., Cooper, W.W., Lewin, A.Y., Seiford, L.M., 1994. Data Envelopment Analysis: Theory, Methodology, and Application. Kluwer, Boston, MA.
Chen, P., 1993. China's challenge to economic orthodoxy: Asian reform as an evolutionary, self-organizing process. China Econ. Rev. 4, 137–142.
Chen, Y., Cook, W.D., Zhu, J., 2010. Deriving the DEA frontier for two-stage processes. Eur. J. Oper. Res. 202 (1), 138–142.
Chen, Y., Cook, W.D., Kao, C., Zhu, J., 2013. Network DEA pitfalls: divisional efficiency and frontier projection under general network structures. Eur. J. Oper. Res. 226 (3), 507–515.
Cherchye, L., Moesen, W., 2004. Institutional Infrastructure and Economic Performance: Levels versus Catching up and Frontier. Katholieke Universiteit Leuven, Leuven.
Climate, inequality, hunger: which global problems would you fix first? [Interactive quiz]. The Guardian.
Coase, R.H., 1977. Economics and contiguous disciplines. In: Perlman, M. (Ed.), The Organization and Retrieval of Economic Knowledge. Westview Press, Boulder, CO, pp. 481–491.
Coase, R.H., 2002. Why Economics Will Change. 4 International Society for New Institutional Economics Newsletter.
Cobb, C., Halstead, T., Rowe, J., 1995. The Genuine Progress Indicator: Summary of Data and Methodology. Redefining Progress. OECD, Washington, DC.
Cook, L., Dimitrov, M., 2017. The social contract revisited: evidence from communist and state capitalist economies. Eur. Asia Stud. 69 (1), 8–26.
Cook, W.D., Liang, L., Zhu, J., 2010. Measuring performance of two-stage network structures by DEA: a review and future perspective. Omega 38, 423–430.
Cooper, W.W., Ijiri, Y. (Eds.), 1983. Kohler's Dictionary for Accountants, sixth ed. Prentice-Hall, New York, NY.
Cooper, W.W., Seiford, L.M., Tone, K., Zhu, J. Some Models and Measures for Evaluating.
Cooper, W.W., Seiford, L.M., Tone, K., 2000. Data Envelopment Analysis: A Comprehensive Text with Models, Applications, References and DEA Solver Software. Kluwer Academic Publishers, Norwell, MA.

Cooper, W.W., Seiford, L.M., Zhu, J., 2004a. Handbook on Data Envelopment Analysis. Kluwer, Boston, MA.
Cooper, W.W., Seiford, L.M., Thanassoulis, E., Zanakis, S.H., 2004b. DEA and its uses in different countries. Eur. J. Oper. Res. 154 (2), 337–344.
Cooper, W.W., Seiford, L.M., Zhu, J., 2005. DEA: past accomplishments and future prospects. J. Prod. Anal. 28 (3), 151–163.
Cooper, W.W., Seiford, L.M., Tone, K., 2006. Introduction to Data Envelopment Analysis and its Uses: With DEA Solver Software and References. Springer, New York, NY.
Cooper, W.W., Seiford, L.M., Tone, K., Zhu, J., 2007. Some models and measures for evaluating performances with DEA: past accomplishments and future prospects. J. Prod. Anal. Springer 28 (3), 151–163.
Cooper, W.W., Thore, S., Tarverdyan, R., 2010. A utility function approach for evaluating country performances with DEA – the twin goals of decent work and a fair globalization. In: Hockley, L.R. (Ed.), Global Operations Management. Nova Science Publishers, New York, NY, pp. 49–72.
Curran, Z., de Renzio, P., 2006. 'What Do We Mean by "Pro-poor Policies" and "Pro-poor Policy Processes"? ODI, London.
Daly, H., Cobb, J., 1989. For the Common Good. Redirecting the Economy toward Community, the Environment and a Sustainable Future. Beacon, Boston, MA.
de Renzio, P., 2005. CSOs and Budgets: Linking Evidence and Pro-poor Policies. ODI, London.
Deardorff, A., Stern, R., 2001. What You Should Know About Globalization and the World Trade Organization. University of Michigan, Ann Arbor, MI.
Debreu, G., 1951. The coefficient of resource utilization. Econometrica 19, 273–292.
Deutsch, K.W., 1966. The Nerves of Government. The Free Press of Glencoe, New York, NY.
Du, J., Chen, Y., Huo, J., 2015. DEA for non-homogenous parallel networks. Omega 56, 122–132.
Eliashberg, J., Winkler, R.L., 1981. Risk sharing and group decision making. Manag. Sci. 27 (11), 1221–1235.
Ellis, H.S., March 1950. The economic way of thinking. Am. Econ. Rev. 1–12.
Emrouznejad, A., Thanassoulis, E., 1996a. An extensive bibliography of data envelopment analysis (DEA). Working Papers (1996) I, 1–55 (Working Paper).
Emrouznejad, A., Thanassoulis, E., 1996b. An extensive bibliography of data envelopment analysis (DEA). Journals Papers (1996) II, 1–21 (Working Paper).
Emrouznejad, A., Thanassoulis, E., 1997. An extensive bibliography of data envelopment analysis (DEA). Supplement III, 1–24 vol. 1. Working Paper 258.
Emrouznejad, A., Yang, G., 2018. A survey and analysis of the first 40 years of scholarly literature in DEA: 1978–2016. Soc. Econ. Plann. Sci. 61.
Emrouznejad, A., Parker, B.R., Tavares, G., 2008. Evaluation of research in efficiency and productivity: a survey and analysis of the first 30 years of scholarly literature in DEA. Soc. Econ. Plann. Sci. 42 (3), 151–157.
Espey, J., Dahmm, H., Manderino, L., 2018. Leaving No U.S. City Behind: The 2018 U.S. Cities SDGs Index. Sustainable Development Solutions Network (SDSN), New York, NY.
Essama-Nssah, B., 2000. New Dimensions of Poverty-Focused Evaluation within a Comprehensive Development Framework. World Bank, Washington, DC.
Esty, D.C., Levy, M., Salteli, A., Dahl, A., 2005. 2005 Environmental Sustainability Index. Yale University with Columbia University, the World Economic Forum, and the Joint Research Centre of the European Commission.

European Commission, 2009. A Roadmap for Action: GDP and Beyond—Measuring Progress in a Changing World. COM (2009) 433 of 2.8.2009.
European Commission, 2012. European Competitiveness Report 2012: Reaping the Benefits of Globalization (ec.europa.eu/enterprise).
European Journal of Operational Research, 1992. 62 (1), 74−84., ISSN 0377-2217.
Färe, R., Grosskopf, S., 1996. Productivity and intermediate products: a frontier approach. Econ. Lett. 50, 65−70.
Färe, R., Grosskopf, S., 2000. Network DEA. Soc. Econ. Plann. Sci. 34, 35−49.
Färe, R., Lovell, C.A.K., 1978. Measuring the technical efficiency of production. J. Econ. Theor. 19 (1), 150−162.
Farrell, M.J., 1957. The measurement of productive efficiency. J. Roy. Stat. Soc. 120 (3), 253−290. JSTOR, www.jstor.org/stable/2343100. Accessed 15 May 2021.
Feldmann, M., Mazepus, H., 2018. State-society relations and the sources of support for the Putin regime: bridging political culture and social contract theory. East Eur. Politics 34 (1), 57−76.
FFF, 1935. An Essay on the Nature and Significance of Economic Science, second ed. Macmillan, London.
FFF, 1938. Live and dead issues in the methodology of economics. Economica 5 (18), 342−352.
FFF, 1955. Economic Analysis, third ed. Harper & Brothers, New York, NY.
FFF, 1969. Economics as a moral science. Am. Econ. Rev. 59 (1), 1−12.
FFF, 1981. Economics and political economy. Am. Econ. Rev. 71 (May), 1−10.
Fiacco, A.V., Kortanek, K.O. (Eds.), 1977. Extremal Methods and Systems Analysis: An International Symposium on the Occasion of Professor Abraham Charnes's Sixtieth Birthday. Springer Verlag, New York, NY.
Flint, R.W., 2013. Practice of sustainable community development. In: Basics of Sustainable Development. Springer, New York, NY, USA; Heidelberg, Germany, pp. 25−54. 33.
Fred, G., 2009. Toshiyuki sueyoshi, contributions of professor William W. Cooper in operations research and management science. Eur. J. Oper. Res. 197 (1), 1−16. ISSN 0377-2217.
Fuchs, H.W., Farrish, R.O.P., Bohall, R.W., 1974. A model of the US apple industry: a quadratic interregional intertemporal activity analysis formulation. Am. J. Agric. Econ. 56, 739−750.
Fukuyama, H., Mirdehghan, S.M., 2012. Identifying the efficiency status in network DEA. Eur. J. Oper. Res. 220, 85−92.
Fukuyama, H., Weber, W.L., 2010. A slacks-based inefficiency measure for a two-stage system with bad outputs. Omega 38 (5), 398−409.
Fukuyama, H., Weber, W.L., Xia, Y., 2015. Time substitution and network effects with an application to Nanobiotechnology Policy for US Universities. Omega. https://doi.org/10.1016/j.omega.2015.04.020.
Gattoufi, S., Oral, M., Reisman, A., 2004a. A taxonomy for data envelopment analysis. Soc. Econ. Plan. Sci. 38 (2−3), 141−158.
Gattoufi, S., Oral, M., Reisman, A., 2004b. Data envelopment analysis literature: a bibliography update (1996−2001). Soc. Econ. Plann. Sci. 38 (2−3), 122−159.
Georgescu-Roegen, N., 1971. The Entropy Law and the Economic Process. Harvard University Press, Cambridge, MA.
Girod, B., van Vuuren, D.P., Hertwich, E.G., 2014. Climate policy through changing consumption choices: options and obstacles for reducing greenhouse gas emissions. Global Environ. Change 25, 5−15.
Goicoechea, A., Hansen, D.R., Duckstein, L., 1982. Multiobjective Decision Analysis with Engineering and Business Applications. Wiley, Hoboken, NJ.

Golany, B., Thore, S., 1997a. The economic and social performance of nations: efficiency and returns to scale. Soc. Econ. Plann. Sci. 31, 191–204.

Golany, B., Thore, S., 1997a. The competitiveness of nations. In: Cooper, W.W., Thore, S., Gibson, D., Phillips, F.F. (Eds.), IMPACT: How IC^2 Research Affects Public Policy and Business Markets. A Volume Honoring George Kozmetsky's Receipt of the National Medal of Technology. Greenwood, Westport, CT, pp. 189–208.

Golany, B., Thore, S., 1997b. Restricted best practice selection in DEA: an overview with a case study evaluating the socio-economic performance of nations. Ann. Oper. Res. 73, 117–140.

Goodland, R., Ledec, G., 1987. Neoclassical Economics and Principles of Sustainable Development. Office of Environmental and Scientific Affairs. The World Bank, Washington, DC, p. 60.

Goreux, L.M., 1981. Independence in Planning: Multi-Level Programming Studies of the Ivory Coast. World Bank, Washington, DC.

Goreux, L.M., Manne, A.S. (Eds.), 1973. Multi-Level Planning: Case Studies in Mexico. North-Holland, Amsterdam.

Guterres, A., July 18, 2020. UN Secretary-General the 18th Nelson Mandela Annual Lecture Delivered Virtually on Nelson Mandela International Day (The theme of the lecture is Tackling the).

Gwartney, J., Wagner, R.E., 1988. The Public Choice Revolution. Intercoll. Rev. 23 (2), 17–26.

Hair, J.F., Anderson, R.E., Tatham, R.L., Black, W.C., 1995. Multivariate Data Analysis. Prentice-Hall, Englewood Cliffs, NJ, p. 53.

Halkos, G.E., Tzeremes, N.G., Kourtzidis, A.A., 2014. A unified classification of two-stage DEA models. Surv. Oper. Res. Manag. Sci. 19, 1–16.

Hawking, S., 1988. A Brief History of Time: From the Big Bang to Black Holes. Bantam Books, New York, NY.

Hazell, P., Poulton, C., Wiggins, S., Dorward, A., 2006. The Future of Small Farms for Poverty Reduction and growth. 2020 Discussion Paper 42, May 2007. IFPRI, Washington, DC.

Hickey, S., King, S., 2016. Understanding social accountability: Politics, power and building new social contracts. J. Dev. Stud. 52 (8), 1225–1240.

Holmstrom, B.R., Tirole, J., 1989. The theory of the firm. In: Schmalensee, R., Willig, R. (Eds.), Handbook of Industrial Organization, Handbook of Industrial Organization, first ed., vol. 1. Elsevier, Amsterdam, pp. 61–133 (chapter 2).

Hueting, R., 1974. New Scarcity and Economic Growth. Dutch ed., Agon (Elsevier: Amsterdam). (1980) Eng. Ed., (North-Holland: Amsterdam).

Humboldt, A.V., 1814–1829. Personal Narrative of Travels to the Equinoctial Regions of the New Continent. vols. 1 and 2, Longman, Hurst, Rees, Orme, and Brown, London.

ILO, 1999. Decent Work. Report of the Director-General. The International Labor Organization, Geneva.

ILO, 2001a. Poverty Reduction and Decent Work in a Globalizing World. The International Labor Organization, Geneva.

ILO, 2001b. Reducing the Decent Work Deficit: A Global Challenge. Report of the Director-General. The International Labor Organization, Geneva.

ILO, 2003. Working Out of Poverty. Report of the Director-General. The International Labor Organization, Geneva.

ILO, 2004. Constitution of the International Labour Organization. Declaration Concerning the Aims and Purposes of the ILO. Geneva 2004. http://www.ilo.org/public/English/about/iloconst.htm.

ILO, 2005. Decent Work and the Millennium Development Goals. Information Folder. The International Labor Organization, Geneva.
ILO, 2008. ILO Declaration on Social Justice for A Fair Globalization, 10 June. The International Labor Organization, Geneva.
ILO, 2019. ILO Declaration for the Future of Work, 26 June. The International Labor Organization, Geneva.
Inter-Agency and Expert Group on Sustainable Development Goal Indicators, 2018, March 6–9. Guidelines on Data Flows and Global Data Reporting for Sustainable Development Goals. UN Statistical Commission.
Ivanova, D., Stadler, K., Steen-Olsen, K., Wood, R., Vita, G., Tukker, A., Hertwich, E.G., 2015. Environmental impact assessment of household consumption. J. Ind. Ecol. 20, 526–536.
Jahanshahloo, G.R., Shirzadi, A., Mirdehghan, S.M., 2008. Finding the reference set of a decision making unit. Asia Pac. J. Oper. Res. 25, 563–573.
Kalbar, P.P., Birkved, M., Hauschild, M., Kabins, S., Nygaard, S.E., 2018. Environmental impact on urban consumption patterns: drivers and focus points. Resour. Conserv. Recycl. 133, 260–269.
Kao, C., 2009. Efficiency decomposition in network data envelopment analysis: a relational model. Eur. J. Oper. Res. 192, 949–962.
Kao, C., 2014a. Efficiency decomposition in network data envelopment analysis with slacks-based measures. Omega 45, 1–6.
Kao, C., 2014b. Network data envelopment analysis: a review. Eur. J. Oper. Res. 239, 1–16.
Kao, C., Hwang, S.N., 2008. Efficiency decomposition in two-stage data envelopment analysis: an application to non-life insurance companies in Taiwan. Eur. J. Oper. Res. 185 (1), 418–429.
Kao, C., Hwang, S.N., 2010. Efficiency measurement for network systems: IT impact on firm performance. Decis. Support Syst. 48 (3), 437–446.
Kaplan, S., 2015. Fixing fragile states: a country-based framework. Norwegian Peacebuilding Resource Centre, Oslo (Report November 2015).
Kharas, H., McArthur, J.W., Ohno, I., 2019. Leave No One behind: Time for Specifics on the Sustainable Development Goals. Brookings Institution Press, Washington, DC (Book available for purchase).
Kimenyi, M.S., 2007. Ten commandments of pro-poor growth. Poverty in Focus 10 (2007), 26–27.
Klein, L.R., 1953. A Textbook of Econometrics. Row, Peterson & Co, White Plains, NY.
Koopmans, T.C. (Ed.), 1951. Activity Analysis of Production and Allocation. Wiley, New York.
Koopmans, Tjalling C., 1975. Concepts of Optimality and Their Uses Nobel Memorial Lecture. Yale University, New Haven, CT.
Krivonozhko, V.E., Førsund, F.R., Lychev, A.V., 2012. Methods for determination of multiple reference sets in the DEA models. J. Math. Stat. 85, 134–138.
Kroll, C., 2015. Sustainable Development Goals: Are the Rich Countries Ready? Bertelsmann Stiftung, Gütersloh.
Krugman, P., 1994. Competitiveness: a dangerous obsession. Foreign Aff. 7, 28–44.
Kuhn, T.S., 1962. The Structure of Scientific Revolutions, 2. University of Chicago Press, Chicago, IL, pp. 177–189. Law, Ethics and Philosophy.
Kuznets, S., 1971. Modern Economic Growth: Findings and Reflections, Nobel Prize Lecture, Nobel Media AB, 2013. Nobelprize.org.
Kynclova, P., Upadhyaya, S., Nice, T., 2020. Composite index as a measure on achieving Sustainable Development Goal 9 (SDG-9) industry-related targets: the SDG-9 index. Appl. Energy 265, 114755.
Lafortune, G., Sachs, J., Schmidt-Traub, G., 2020. It's Called the Sustainable Development Goals Index for a Reason. Foreign Policy. (In Response to This Opinion piece).

Lamy, P., 2006. WTO Diretor General Speech, "Humanising Globalization". Chilem, Santiago de Chile.
Land, K.C., Lovell, C.A.K., Thore, S., 1992. Productive efficiency under capitalism and state socialism: the chance-constrained programming approach. Publ. Finance 47, 109–121 (Supplement: Public Finance in a World of Transition ed. by P. Pestieau).
Land, K.C., Lovell, C.A.K., Thore, S., 1993. Chance-constrained data envelopment analysis. Manag. Decis. Econ. 14, 541–554.
Land, K.C., Lovell, C.A.K., Thore, S., 1994. Productive efficiency under capitalism and state socialism: an empirical inquiry using chance constrained data envelopment analysis. Tech. Fore. Soc. Change 46, 139–152.
Lawn, P.A., Sanders, R.D., 1999. Has Australia surpassed its optimal macroeconomic scale? Finding out with the aid of 'benefit' and 'cost' accounts and a sustainable net benefit index. Ecol. Econ. 28 (2), 213–229.
Leontief, W., Carter, A.P., Petri, P., 1977. The Future of the World Economy. A United Nations Study. Oxford University, New York, NY.
Levi-Faur, D. (Ed.), 2012. The Oxford Handbook of Governance. Oxford University, Cambridge, MA.
Lewis, H.F., Sexton, T.R., 2004. Network DEA: efficiency analysis of organizations with complex internal structure. Comput. Oper. Res. 31, 1365–1410.
Liang, L., Cook, W.D., Zhu, J., 2008. DEA models for two-stage processes: game approach and efficiency decomposition. Nav. Res. Logist. 55, 643–653.
Liobikiene, G., Mandravickait ė, J., 2013. Convergence of new members of the EU: changes in household ˙ consumption expenditure structure regarding environmental impact during the prosperous period. Environ. Dev. Sustain. 15, 407–427.
Littlechild, S.C., Thompson, G.F., 1977. Aircraft landing fees: a game theory approach. Bell J. Econ. 8, 186–204.
Loewe, M., Zintl, T., Houdret, A., 2021. The social contract as a tool of analysis: Introduction to the special issue on "Framing the evolution of new social contracts in Middle Eastern and North African countries". World Dev. 145, 104982.
Loh, J., Green, R.E., Ricketts, T., Lamoreux, J., Jenkins, M., Kapos, V., Randers, J., 2005. The Living Planet Index: using species population time series to track trends in biodiversity. Phil. Trans. R. Soc. B. 360, 289–295.
Lovell, K., 1995. Measuring the macroeconomic performance of the Taiwanese economy. Int. J. Prod. Econ. 39 (1–2), 165–178.
Lozano, S., 2011. Scale and cost efficiency analysis of networks of processes. Expert Syst. Appl. 38 (6), 6612–6617.
Lynch, A., LoPresti, A., Fox, C., 2019. The 2019 US Cities Sustainable Development Report. Sustainable Development Solutions Network (SDSN), New York, NY.
Malthus, T., 1798. An Essay of the Principle of Population. Johnson, London.
Maragos, E., Despotis, D., 2002. Comparing, multiobjective mathematical programming methods in the light of data envelopment analysis. J. Interdiscipl. Math. 5 (3), 221–230.
Marks, N., 2010. Happy Planet Index. www.happyplanetindex.org.
Meadows, D., 1972. The Limits to Growth. A Report for the Club of Rome. Universe Books, New York, NY.
Meadows, D.H., Meadows, D.L., Randers, J., Behrens III, W.W., 1972. The Limits to Growth. New American Library, New York, NY.
Meadows, D.H., Meadows, D.L., Randers, J., 1992. Beyond the Limits: Confronting Global Collapse, Envisioning a Sustainable Future. Chelsea Green Publishing, Post Mills, VT.
Mehdiloozad, M., Mirdehghan, S.M., Sahoo, B.K., Roshdi, I., 2015. On the identification of the global reference set in data envelopment analysis. Eur. J. Oper. Res. 245, 779–788.

Moyer, J.D., Hedden, S., 2020. Are we on the right path to achieve the sustainable development goals? World Dev. 127, 104749, 32.

NADEL, 2007. Working with a Sustainable Livelihoods Approach. NADEL, Zurich, and SDC, Bern.

Nature Research & TReNDS, 2020. Putting data to work for real-world SDG progress. Nature. https://www.nature.com/articles/d42473-020-00334-2.

Negishi, T., 1962. The stability of a competitive economy: a survey article. Econometrica 639.

Nicolis, G., Prigogine, I., 1977. Self-Organization in Nonequilibrium Systems: From Dissipative Structures to Order Through Fluctuations. Wiley, New York, NY.

Nordhaus, W., Tobin, J., 1972. Is Growth Obsolete? Columbia University, New York, NY.

Nussbaum, M.C., 2000. Women and Human Development: The Capabilities Approach. Cambridge University Press, Cambridge, MA.

Nussbaum, M.C., Sen, A. (Eds.), 1993. The Quality of Life. Clarendon Press, Oxford.

OECD, 2001a. The DAC Guidelines: Poverty Reduction. OECD, Paris.

OECD, 2001b. Strategies for Sustainable Development, the DAC Guidelines. OECD Publishing, Paris. https://doi.org/10.1787/9789264194762-en.

OECD, 2004. Measuring Sustainable Development; Integrated Economic, Environmental and Social Frameworks. OECD Publishing, Paris, pp. 39–52.

OECD, 2005. Paris Declaration on Aid Effectiveness. OECD, Paris.

OECD, 2006. Promoting Pro-poor Growth: Key Policy Messages. OECD, Paris.

OECD, 2007. Promoting Pro-poor Growth: A Practical Guide to Ex-Ante Poverty Impact Assessment. OECD, Paris.

OECD, 2008. The OCED/JRC Handbook on Constructing Composite Indicators Identifies Ten Steps for Building an Index. (See Table 1 for More detail.) in Short, However, These 10 Steps Cover the Following. OECD, Paris.

OECD, 2009. Concepts and dilemmas of state-building in fragile situations: from fragility to resilience. OECD J. Dev. 9 (3), 61–148.

OECD, 2017. Trust and Public Policy: How Better Governance Can Help Rebuild Public Trust. OECD Public Governance Reviews, Paris.

OECD, 2020. Measuring Distance to the SDG Targets; an Assessment of where OECD Countries Stand, 2017. Available online. http://www.oecd.org/sdd/OECD-Measuring-Distance-to-SDG-Targets.pdf (Accessed 20 December 2020).

OECD, C/MIN, 2005. Statement to the Follow-Up of the UN Millennium Declaration and Monterey Consensus.

OECD/EC JRC, 2008. Handbook on Constructing Composite Indicators, Methodology, and User Guide. OECD Publishing, Paris, ISBN 978-92-64-04345-9.

OMCFG, 2013. Now for the Long Term. Cambridge, Oxford.

OMNI Magazine, 1983. Ilya Prigogine: Wizard of Time. Interview by R.B. Tucker, May 1983 at. www.edu365.cat.

Osberg, L., Sharpe, L., 2002. An index of economic well-being for selected OECD countries. Rev. Income Wealth 48, 291–316.

Owen, G., 1982. Game Theory, second ed. Academic, New York, NY. Now in 4th ed. (2013).

Papadimitriou, E., Neves, A.R., Becker, W., 2019. Behind the Numbers: Joint Research Centre Audit of the SDG Index and Dashboards.

Pareto, V., 1909. Manuel D'économie Politique. Droz, Paris, pp. 32–38. Chapter VI, sections.

Pastor, J.T., Ruiz, J.L., Sirvent, I., 1999. An enhanced Russell graph efficiency measure. Eur. J. Oper. Res. 115, 596–607.

Pearce, D.W., 1980. The MIT Dictionary of Modern Economics. The MIT Press, Cambridge, MA.
Pearce, D.W., Atkinson, G., 1993. Capital theory and the measurement of sustainable development: an indicator of weak sustainability. Ecol. Econ. 8, 103–108.
Piketty, T., 1997. L'Économie des Inégalités (the economics of inequality). Editions la Découverte, Paris.
Piketty, T., 2013a. Capital in the Twenty- First Century. Harvard University Press.
Piketty, T., 2013b. Le Capital au XXIe siècle. Editions du Seuil, coll, Paris. « Les Livres du monde ».
Poincaré Henri in Stanford Encyclopedia of Philosophy http://plato.stanford.edu/entries/poincare/.
Poom, A., Ahas, R., 2016. How does the environmental load of household consumption depend on residential location? Sustainability 8, 799.
Porter, M., 1980. Competitive Strategy. Free Press, New York, NY.
Porter, M., 1985. Competitive Advantage. Free Press, New York, NY.
Porter, M., 1990. The Competitive Advantage of Nations. Harvard Business Review (March, April). http://www.economie.ens.fr/IMG/pdf/porter_1990_-_the_competitive_advantage_of_nations.pdf.
Prieto, A.M., Zofio, J.L., 2001. Evaluating effectiveness in public provision of infrastructure and equipment: the case of Spanish municipalities. J. Prod. Anal. 15, 41–58.
Prieto, A.M., Zofio, J.L., 2007. Network DEA efficiency in input-output models: with an application to OECD countries. Eur. J. Oper. Res. 178, 292–304, 2007.
Prigogine, I., 1997. The End of Certainty: Time, Chaos, and the New Laws of Nature. Free Press, New York, NY.
Prigogine, I., Stengers, I., 1984. Order Out of Chaos. Bantam, New York, NY.
Public Choice Revolution, 1988. The. Intercollegiate Review, 23, pp. 17–26 no. 2.
Rees, W.E., 1992. Ecological footprints and appropriated carrying capacity: what urban economics leaves out. Environ. Urban. 4 (2), 121–130.
Regeni, G., Auktor, G., 2017. The Developmental State in the 21st Century: Calling for a New Social Contract. German Development Institute/Deutsches Institut für Entwicklungspolitik, Bonn (Discussion Paper 5/2017).
Rezaee, M.J., Izadbakhsh, H., Yousefi, S., 2016. An improvement approach based on DEA-game theory for comparison of operational and spatial efficiencies in urban transportation systems. KSCE J. Civ. Eng. 20, 1526–1531.
Rhodes, M., Mény, Y. (Eds.), 2016. The Future of European Welfare: A New Social Contract?. Palgrave McMillan, Houndsmills, NY.
Robbins, L.C., 1932. An Essay on the Nature and Significance of Economic Science. Macmillan, London.
Rousseau, J.-J., 1762. Du contrat social ou principes du droit politique.
Sachs, J., Schmidt-Traub, G., Kroll, C., Lafortune, G., Fuller, G., Sustainable Development Report, 2019. Bertelsmann Stiftung: NGütersloh, Germany. Sustainable Development Solutions Network (SDSN), New York, NY.
Samuelson, P.A., 1952. Spatial price equilibrium and linear programming. Am. Econ. Rev. 42, 283–303.
Samuelson, P.A., 1954. The pure theory of public expenditure. Rev. Econ. Stat. 36 (4), 387–389. http://links.jstor.org/sici?sici=0034-6535%28195411%2936%3A4%3C387%3ATPTOPE%3E2.0.CO%3B2-A.
SCC, 2018. SCC (Social Contract Center) (2018): Social Contract Center: A Decent Life for All Egyptians. SCC, Cairo. Available from: http://www.socialcontract.gov.eg/English/About-Us (Accessed 27 August 2019).

Schaible, S., 1996. Fractional programming. In: Gass, S.I., Harris, C.M. (Eds.), Encyclopedia of Operations Research and Management Science. Kluwer Academic Publishers, Norwell, MA.

Schmidt-Traub, G., Hoff, H., Bernlöhr, M., 2019. Policy Brief: International Spillovers and the Sustainable Development Goals. Sustainable Development Solutions Network, New York, NY.

SDC, 2004. Creating the Prospects of Living a Life in Dignity. SDC, Bern.

SDSN TReNDS, 2020. Leaving No One off the Map: A Guide for Gridded Population Data for Sustainable Development (Available in English, Spanish, and French).

Seiford, L.M., 1994. A DEA bibliography (1978–1992). In: Charnes, Cooper, W.W., Lewin, A.Y., Seiford, L.M. (Eds.), Data Envelopment Analysis: Theory, Methodology and Applications. Kluwer, Boston, MA, pp. 437–469.

Seiford, L.M., 1997. A bibliography for data envelopment analysis (1978–1996). Ann. Oper. Res. 73, 393–438.

Seiford, L.M., Thrall, R.M., 1990. Recent developments in DEA: the mathematical programming approach to frontier analysis. J. Econom. 4, 7–38.

Sen, A.K., 1985. Commodities and Capabilities. Oxford University Press, Oxford.

Sen, A.K., 1999. Development as Freedom. Knopf, New York, NY.

Serra, N., Stiglitz, J.E. (Eds.), 2008. The Washington Consensus Reconsidered: Towards a New Global Governance. Oxford University Press, New York, NY.

Sexton, T.R., Lewis, H.F., 2003. Two-stage DEA: an application to major league baseball. J. Prod. Anal. 19, 227–249.

Simon, H.A., 1978. Rational Decision-Making In Business Organizations, Nobel Memorial Lecture. Carnegie-Mellon University *, Pittsburgh, PA.

Stiglitz, J.E., Sen, A.S., Fitoussi, J.-P., 2009. Report by the Commission on the Measurement of Economic Performance and Social Progress. CMEPSP.

Stone, P., 2014. Social Contract Theory in the Global Context. Law Ethics Philos. 2, 177–189 [online]. https://raco.cat/index.php/LEAP/article/view/297565.

Swiss Federal Palace, Swiss Neutrality, 4th Revised Edition Communication DDPS, Federal Palace East Wing, 3003 Bern.

Takayama, T., Judge, G.G., 1971. Spatial and Temporal Price and Allocation Models. North-Holland, Amsterdam.

Tarverdyan, R., 2012. Geneva Consensus For Sustainable Development: Toward a New Social Contract. UN Special N^0 721.

Tavares, G., 2002. A Bibliography of Data Envelopment Analysis (1978–2001). RUTCOR, Rutgers University, New Brunswick, NJ. Available from: http://rutcor.rutgers.edu/pub/rrr/reports2002/1_2002.pdf.

Terleckyj, N., 1975. Improvements in the Quality of Life 1974-1983: Estimates of Possibilities in the United States. National Planning Association, Washington, DC.

The World Commission on Environment and Development's, 1987. The Brundtland Commission Report Our Common Future. Oxford University Press, Oxford.

Thore, S., 1956. A Critique of the Theory of the Consumption Function (Inst. Univ. d'Études Européennes: Turin). With a Foreword by H. Wold.

Thore, S., 1964. Price strategies of an 'ignorant' monopolist. In: Wold, H. (Ed.), Econometric Model Building. North-Holland, Amsterdam, pp. 283–296.

Thore, S., 1986. Generalized network spatial equilibrium: the deterministic and the chance-constrained case. Pap. Reg. Sci. Assoc. Reg. Sci. Assoc. Meet. 59, 93–102.

Thore, S., 1987. Chance-constrained activity analysis. Eur. J. Oper. Res. 30, 267–269.

Thore, S., 1991. Economic Logistics: The Optimization of Spatial and Sectoral Resource, Production, and Distribution Systems. With a Chapter on "Some Elements of Saddle-point Theory" Coauthored by W.W. Cooper and S. Thore. Quorum, Westport, CT.

Thore, S., 1995. The Diversity, Complexity, and Evolution of High Tech Capitalism. Kluwer, Boston, MA.
Thore, S., 1996. Economies of scale in the US computer industry: an empirical investigation using data envelopment analysis. J. Evol. Econ. 6, 199–216.
Thore, S., 1999. Enterprise in the information age. In: Halal, W.E., Taylor, K.B. (Eds.), Twenty-First Century Economics. St. Martin's, New York, NY.
Thore, S. (Ed.), 2002. Technology Commercialization: DEA and Related Analytical Methods for Evaluating the Use and Implementation of Technical Innovation, 2nd printing. Kluwer, Boston, MA. Springer 2012.
Thore, S., 2003. Global optimization of economic and social policy. In: Wang, J. (Ed.), Encyclopedia of Business Analytics and Optimization (in press) 2014.
Thore, S., 2014. Global optimization of economic and social policy. In: Encyclopedia of Business Analytics and Optimization. https://doi.org/10.4018/978-1-4666-5202-6.ch099.
Thore, S., Tarverdyan, R., 2008. Using data envelopment analysis to screen the possibility of A Fair Globalization. Port. J. Manag. Stud. 13, 129–151.
Thore, S., Tarverdyan, R., 2009. Using data envelopment analysis to quantify ILO objectives and identify policies conducive to Decent Work in a globalized world. Soc. Econ. Plann. Sci. 43, 151–164.
Thore, S., Tarverdyan, R., 2015. Diagnostics for a Globalized World, vol. 7. World Scientific-Now Publishers Series In Business. http://www.worldscientific.com/worldscibooks/10.1142/9421.
Thore, S., Tarverdyan, R., 2016. The sustainable competitiveness of nations. In: Technol. Forecast Soc. Chang., 106, pp. 108–114.
Thore, S., Golany, B., Tarverdyan, R., Adler, N., Yazhemsky, E., 2015. Measuring the SDGs Progress with DEA; Brief for GSDR 2015. https://sustainabledevelopment.un.org/content/documents/6729140-Thore-Measuring%20the%20SDGs%20Progress%20with%20DEA.pdf.
Tinbergen, J., 1987. Revitalizing the United Nations System, Waging Peace Series, Booklet 13. Santa Barbara, CA (1187 Coast Village Road, Suite 123, Santa Barbara CA 93108). Nuclear Age Peace Foundation.
Tinbergen, J., 1939. Statistical Testing of Business Cycle Theories, 2 vols. League of Nations, Geneva.
Tinbergen, J., 1941. Econométrie. Noorduijn, Gorinchem.
Tinbergen, J., 1952. On the Theory of Economic Policy. North-Holland, Amsterdam.
Tinbergen, J., 1956. Economic Policy: Principles and Design. North-Holland, Amsterdam. http://repub.eur.nl/resource/pub_16740/index.html.
Tinbergen, J., 1969. The use of models: experience and prospects. In: Lindbeck, A. (Ed.), Nobel Lectures, Economics 1969–1980. World Scientific Publishing, Singapore, 1992.
Tinbergen, J., 1987. Revitalizing the united nations system. In: Waging Peace Series, Booklet 13. Nuclear Age Peace Foundation, Santa Barbara, CA.
Tone, K., Tsutsui, M., 2014. Dynamic DEA with network structure: a slacks-based measure approach. Omega 42 (1), 124–131.
Tone, K., Tsutsui, M., 2009. Network DEA: A slacks-based measure approach". Eur. J. Oper. Res. 197 (1), 243–252.
Toward a New Social Contract: Taking on Distributional Tensions in Europe and Central Asia.
Townsend, P., 1979. Poverty in the United Kingdom. Penguin, Harmondsworth.
UFC, 2004. The Barcelona Agenda. Universal Forum of Cultures, Barcelona.
UN Department of Economic and Social Affairs. Sustainable Development Goal 17. Available online: https://sdgs.un.org/goals/goal17 (Accessed 20 December 2020).

UN, 1948. Universal Declaration of Human Rights, of the General Assembly. United Nations, New York, NY.
UN, 1966. Int'l Covenant of Economic, Social, and Cultural Rights, of the General Assembly. United Nations, New York, NY.
UN, 1992. Report of the United Nations Conference on Environment and Development Rio de Janeiro, 3-14 June 1992. https://www.un.org/esa/dsd/agenda21/Agenda%2021.pdf.
UN, 1995. Report of the World Summit for Social Development. United Nations, Copenhagen.
UN, 2000a. Assessing the Social Impact of Economic Measures Taken in Response to Financial Crises in Developing Countries, Report of the Sec'y-Gen'l to the Special Session of the General Assembly. United Nations, Geneva.
UN, 2000b. United Nations Millennium Declaration. United Nations, New York, NY.
UN, 2001a. Enhancing Social Protection and Reducing Vulnerability in a Globalizing World, Report of the Sec'y-Gen'l, Economic and Social Council. United Nations, Geneva.
UN, 2001b. Road Map toward the Implementation of the United Nations Millennium Declaration. United Nations, New York, NY.
UN, 2002a. Deepening Democracy in a Fragmented World. United Nations, Geneva.
UN, 2002b. Political Declaration. World Summit on Sustainable Dev't, Johannesburg. United Nations, Geneva.
UN, 2004. UN Department of Economics and Social Affairs, Division for Sustainable Development, Monterrey Consensus of the International Conference on Financing for Development. United Nations, New York, NY. March 22, 2002.
UN, 2005. 2005 World Summit Outcome, Resolution Adopted by the General Assembly. United Nations, New York, NY.
UN, 2006. UN Economic and Social Council, Creating an Environment at the National and International Level Conducive to Generating Full and Productive Employment and Decent Work for All. Report of the Secretary General. United Nations, Geneva.
UN, 2012a. Millennium Development Goals Indicators. United Nations, Geneva.
UN, 2012b. The Future We Want, of the Rio +20 Conference. United Nations, Geneva.
UN, 2013. A New Global Partnership: Eradicate Poverty and Transform Economies Through Sustainable Development, Report of the High Level Panel of Eminent Persons on the Post-2015 Development Agenda. United Nations, Geneva.
UN Development Programme. reportHuman Development Report 1990: Concept and Measurement of Human Development. Available online: http://hdr.undp.org/en/reports/global/hdr1990 (Accessed 20 December 2020).
UN ECOSOC, 2020. Progress towards the Sustainable Development Goals: Report of the Secretary-General (Also Available in Arabic, Chinese, French, Russian, and Spanish).
UN ECOSOC, 2020. Report of the Food and Agriculture Organization of the United Nations on Recent Developments in Agricultural and Rural Statistics.
UNDP (United Nations Development Programme), 1990. HDR 1990 - Concept and Measurement of Human Development. Human Development Report Office (HDRO).
UNDP (United Nations Development Programme), 2016. Local Governance in Fragile and Conflict-Affected Settings. Building a Resilient Foundation for Peace and Development, New York, NY.
UNDP, 2019. Leaving No One behind: Towards Inclusive Citizenship in Arab Countries. Arab Human Development Report United Nations Development Programme, New York, NY (2019).
UNDP and AFESD, 2002. UNDP (united nations development programme), & AFESD (united nations development programme/arab fund for economic and social

development). In: Creating Opportunities for Future Generations. Arab Human Development Report 2002, New York, NY (2002).
UNDP and NOREF, 2016. Engaged Societies, Responsive States. UNDP (United Nations Development Programme), & NOREF (Norwegian Peacebuilding Resource Center) (2016). The Social Contract in Situations of Conflict and Fragility, New York and Oslo.
UNFPA (United Nations Population Fund), 2017. State of World Population 2017 – Worlds Apart: Reproductive Health and Rights in an Age of Inequality. UNFPA, New York, NY.
United Nation, 2019. SDG Summit 2019; Summary of the President of the General Assembly. UN General Assembly, New York, NY.
United Nations, 2006. Delivering as One, Report of the Secretary-General's High-Level Panel on UN System-wide Coherence. United Nations, New York, NY.
United Nations, 2009. Report of the Commission of Experts of the President of the United Nations General Assembly on Reforms of the International Monetary and Financial System. 21 September. United Nations, New York, NY. http://www.un.org/ga/econcrisissummit/docs/FinalReport_CoE.pdf. Accessed 19 January 2012.
United Nations, 2013a. A New Global Partnership: Eradicating Poverty and Transform Economies through Sustainable Development. United Nations Publications, New York, NY.
United Nations, 2013b. A New Global Partnership: Eradicating Poverty and Transform Economies through Sustainable Development, the Report of the High-Level Panel of Eminent Persons on the Post-2015 Development Agenda. http://www.un.org/en/development/desa/policy/untaskteam_undf/HLP%20P2015%20Report.pdf.
United Nations, 2013c. MDG Gap Task Force Report 2013 – the Global Partnership for Development: The Challenge We face. Sales No. E.13.I.5.
United Nations, 2013d. A Life of Dignity for All: Accelerating Progress towards the Millennium Development Goals and Advancing the United Nations Development Agenda beyond 2015 Report of the Secretary-General. Available from: http://www.un.org/millenniumgoals/pdf/A%20Life%20of%20Dignity%20for%20All.pdf.
United Nations, 2014a. A World that Counts. Mobilizing the Data Revolution for Sustainable Development, Report Prepared at the Request of the United Nations Secretary-General, by the Independent Expert Advisory Group on a Data Revolution for Sustainable Development. Available from: http://www.undatarevolution.org/wp-content/uploads/2014/11/A-World-That-Counts.pdf.
United Nations, 2014b. Prototype Global Sustainable Development Report. United Nations Department of Economic and Social Affairs, Division for Sustainable Development, New York, NY. http://sustainabledevelopment.un.org/globalsdreport/23.
United Nations, 2015a. Transforming Our World: The 2030 Agenda for Sustainable Development, United Nations General Assembly Resolution 70/1, 25 September 2015. Available from: https://sustainabledevelopment.un.org/post2015/transformingourworld.
United Nations, 2015b. UN General Assembly 2015, Transforming Our World: The 2030 Agenda for Sustainable Development, A/RES/70/1 (21 October). www.refworld.org/docid/57b6e3e44.html. Accessed 3 August 2017.
United Nations, 2015c. Global Sustainable Development Report. United Nations Department of Economic and Social Affairs, New York, NY.
United Nations, 2015d. Millennium Development Goals and beyond 2015. Available from. www.un.org/millenniumgoals.
United Nations Commission on Sustainable Development (UNCSD) Secretariat, 2011a. IFSD: Issues Related to an Intergovernmental Body on SD. RIO 2012 Issue Briefs No 3 (October). http://www.uncsd2012.org/rio20/content/documents/209Issues%20Brief%203%20 - %20IFSD%20SD%20Body_final_final_111111.pdf. Accessed 19 January 2012.

United Nations Commission on Sustainable Development (UNCSD) Secretariat, 2011b. Regional, National and Local Level Governance for Sustainable Development. RIO 2012 Issue Briefs No 10 (December). www.uncsd2012.org/rio20/content/documents/229draft_15Dec7 pm_formatted730am.pdf. Accessed 19 January 2012.

United Nations, General Assembly, 2012. The Future We Want. Resolution 66/288 of 27 July 2012. Available from. http://www.un.org/ga/search/view_doc.asp?symbol=A/RES/66/288&Lang=E.

United Nations Research Institute for Social Development (UNRISD), 2011. Green Economy and Sustainable Development: Bringing Back the Social. UNRISD Submission to the UNCSD, 1 November. www.unrisd.org/80256B42004CCC77/(httpInfoFiles)/44AC8E087556E45AC12579 3D0056E4B7/$file/UNCSD%20input_UNRISD.pdf. Accessed 19 January 2012.

United Nations University Institute of Advanced Studies (UNU/IAS), 2002. International Sustainable Development Governance: The Question of Reform: Key Issues and Proposals. UNU/IAS Final Report, August 2002, Prepared for the World Summit on Sustainable Development. http://www.uncsd2012.org/rio20/content/documents/ISDGFinalReport.pdf. Accessed 19 January 2012.

UNSD, 2020a. E-handbook on Sustainable Development Goals Indicators.

UNSD, 2020b. The Sustainable Development Goals Report.

UNWCED, 1987. Our Common Future Aka the Brundtland Report, by UN World Commission on Environment & Dev't. Oxford University, Cambridge, MA.

USCC, 2006. The Competitiveness Index: Where the US Stands. US Council of Competitiveness. www.compete.org.

USCC, 2013. Making Impact, Annual Report 2012–2013. US Council of Competitiveness.

van de Kerk, G., Manuel, A.R., 2008. A comprehensive index for a sustainable society: the SSI— the Sustainable Society Index. Ecol. Econ. 66 (2–3), 228–242.

von Neumann, J., Morgenstern, O., 1947. Theory of Games and Economic Behavior. Princeton University, Princeton, NJ. Now available in a 60th Anniversary Commemorative Edition (2007).

Wackernagel, M., 1994. Ecological Footprint and Appropriated Carrying Capacity: A Tool for Planning toward Sustainability (PhD thesis). School of Community and Regional Planning. The University of British Columbia, Vancouver, BC.

Waldrop, M.M., 1992. Complexity: The Emerging Science at the Edge of Order and Chaos. Simon & Schuster, New York, NY.

Wang, C.H., Gopal, R., Zionts, S. Use of Data Envelopment Analysis in Assessing Information Technology impact on firm performance. Ann. Oper. Res. 73, 191–213. https://doi.org/10.1023/A:1018977111455.

World Bank, 2019. © World Bank. Washington, DC. https://openknowledge.worldbank.org/handle/10986/30393. License: CC BY 3.0 IGO, Washington, DC.

WCSDG, 2004a. A Fair Globalization: Creating Opportunities for All. World Commission.

WCSDG, 2004b. The Social Dimension of Globalization. World Commission.

WEF, 2011. The Global Competitiveness Report 2011–2012, Geneva 2011.

Werner, S. (Ed.), 1989. The State of Economic Science: Views of Six Nobel Laureates. W.E. Upjohn Institute for Employment Research, Kalamazoo, MI. https://doi.org/10.17848/9780880995962.

Wiener, N., 1950. The human use of human beings: cybernetics and society. Houghton Mifflin.

Wiener, N., 1961. Cybernetics, or Control and Communication in the Animal and the Machine, second ed. MIT Press; John Wiley & Sons Inc.

Wiener, N., 1989. The Human Use of Human Beings: Cybernetics and Society. Free Association Books, London.
Williamson, J. (Ed.), 1990. Latin American Adjustment: How Much Has Happened?. IIE, Washington, DC.
Williamson, J., 2008. A short history of the Washington consensus. In: Serra, N., Stiglitz, J.E. (Eds.), The Washington Consensus Reconsidered: Towards a New Global Governance. Oxford University, New York, NY.
Williamson, J., Wold, H., 1964. Econometric Model Building. Essays on the Causal Chain Approach. North-Holland, Amsterdam.
Wold, H., Jureen, L., 1952. Demand Analysis. A Study in Econometrics. Almquist & Wiksell, Uppsala.
World Bank, 1998. The Social Development Family in the Environmentally and Socially Sustainable Network. World Bank, Washington, DC.
World Bank, 2008. Annual Review of Development Effectiveness 2008: Shared Global Challenges. Independent Evaluation Group, Washington, DC.
World Bank, 2015a. Towards a New Social Contract. World Bank, Washington, DC. Retrieved.
World Bank and IMF, 2004. Policies and Actions for Achieving the MDGs and Related Outcomes, Global Monitoring Report. World Bank, Washington, DC.
World Bank and IMF, 2006a. A Common Vision for the World Bank Group. World Bank and IMF, Washington, DC.
World Bank and IMF, 2006b. Millennium Development Goals: Strengthening Mutual Accountability, Aid, Trade, and Governance, Global Monitoring Report. World Bank, Washington, DC.
World Bank Group, 1998. Partnerships for Development: Proposed Action for the World Bank. World Bank Group, Washington, DC.
World Economic Forum, 2018. The Inclusive Development Index 2018: Summary and Data Highlights released January 22.
Xu, Z., Chau, S.N., Chen, X., et al., 2020. Assessing progress towards sustainable development over space and time. Nature 577, 74–78. https://doi.org/10.1038/s41586-019-1846-3.
Yorke, J., Li, T.–Y., 1975. Period three implies chaos. Am. Math. Mon. 82, 985–992.
Zahedi, F., 1986. Group consensus function estimation when preferences are uncertain. Oper. Res. 34 (6), 883–894.

Index

'*Note*: Page numbers followed by *f* indicate figures and *t* indicate tables.'

FSC
www.fsc.org
MIX
Papier | Fördert
gute Waldnutzung
FSC® C083411

Kolibri 360 GmbH
Ferdinand-Jühlke-Straße 7
99095 Erfurt, Deutschland
produktsicherheit@kolibri360.de